Mocking Bird Technologies

Mocking Bird Technologies

The Poetics of Parroting, Mimicry, and Other Starling Tropes

Christopher GoGwilt and Melanie D. Holm

Editors

FORDHAM UNIVERSITY PRESS

New York 2018

Fordham University Press also publishes its books in a variety of electronic formats. Some content that appears in print may not be available in electronic books.

Visit us online at www.fordhampress.com.

Library of Congress Cataloging-in-Publication Data available online at https://catalog.loc.gov.

Printed in the United States of America

20 19 18 5 4 3 2 1

First edition

CONTENTS

PREFACE

This volume offers an extended meditation on bird mimicry. In a set of original essays on the comparative and global poetics of bird mimicry, illustrated by verbal and visual specimens, the collection embraces a range of theoretical and critical perspectives. The essays examine the historical, poetic, and semiotic problem of mimesis exemplified both by the imitative behavior of starlings, parrots, and other mocking birds and by the poetic trope of such birds in a range of literary and philological traditions. The volume has its origin in a series of talks and discussions organized for the 2013 meeting of the American Comparative Literature Association. The theme of the ACLA conference—"Global Positioning Systems"—playfully highlighted the increasing emphasis in literary and cultural studies on technology and on the "global," as well as the interrelation between the two. At the heart of this theme lies a challenge to traditional comparative literature. With its disciplinary origins in nineteenth-century European comparative philology, traditional comparative literature aligned peoples with languages, and languages with nations. Contemporary "global" studies, by contrast, emphasize processes of migration, mobility, and circulation while turning away from human language to consider systems or networks across oceans, seas, and deserts, rather than between nations, continents, and civilizations. In responding to the new "global positioning" of comparative literature as a discipline, the panel discovered a literary *topos* that simultaneously embodies the new "global" turn and reaches back to the roots of the old comparative and linguistic disciplinary model. In further developing this topic to produce the current volume, we have sought to emphasize neither one nor the other of these poles—neither the bird's-eye view of the global nor the area-specific poetics of literary birdsong—but rather to do justice to both by exploring their overlapping and interrelated relevance. Taken together, the essays offer new models for combining comparative and global studies of literature and culture.

Mocking Bird Technologies focuses on techniques of art, artifice, and paralinguistic performance in literary theory and practice. Each of the

INTRODUCTION

Parrots and Starlings

Christopher GoGwilt and Melanie D. Holm

Mocking Bird Technologies brings together a range of different essays on the poetics of bird mimicry with the overall aim of challenging the comparative and global premises of contemporary literary studies. Each of the chapters in this volume addresses the mimicry of birds in literature, whether this entails the ways birds mock humans, the ways humans mock birds, or some other permutation of interspecies mocking, mimicry, or imitation. The technologies of bird mimicry, in all these senses, are a foundational feature of poetics. The "mocking birds" of our title do not primarily refer to the New World group of birds (of the Mimidae family), although mockingbirds in their biodiversity present an exemplary problem of both scientific and poetic classification. The striking variety of mockingbirds Darwin found on the Galápagos Islands helped precipitate doubts about "the stability of species" that pushed him to theorize the principles of evolution.[1] Rather than referring to any one kind of bird, however, our title's emphasis on "mocking bird technologies" is intended to draw attention to many kinds of birds and to the fullest possible range of interspecies "mocking" or mimicry. Most of the essays focus on a particular kind of bird in literature: parrot, starling, dove, pigeon, goldfinch, lark, seagull. While each essay brings a different combination of theoretical, critical, and methodological approaches to bear on the topic, all of them, in one way or another, address how bird mimicry constitutes a foundational feature of literature (in theory and in practice) considered from both comparative and global perspectives.

The structure of the volume embraces the technologies of bird mimicry with an experimental way of reading that seeks to emphasize the inherently ekphrastic form of bird mimicry itself. The "technologies" of our title refer to the techniques birds use to mimic and to the techniques of poetry often associated with bird forms (and in many ancient traditions). "Technology" suggests the ancient Greek *techne* (which Heidegger famously connects with *poeisis* even as he questions the modern sense of "technology"). The "technologies" of our title are also intended to cover a broader range of possibilities, including critical and creative modes of presentation, and extending across various media and even beyond (or alongside) language itself. Following the format of a collection of critical essays, thematically and chronologically arranged, the volume presents the essays as critical reflections on instances of mocking bird technologies, surrounded by sample specimens. While critical essays comprise the main substance of the collection, each essay is prefaced by a mock précis composed in what GoGwilt in the Afterword calls "starling" form (composed in three columns of twenty-seven words going down from right to left). These starling summaries are designed to highlight a disruption of critical presumptions at work in the essays themselves. The interplay of theoretical argument and creative response is part of the book's extended meditation on the poetics of bird mimicry, defying stasis and definition and signaling an imitative and improvisational dynamic of reading and interpretation, text and critic. Expanding the traditional analytic presentation of the essays, the volume includes a set of illustrative specimens of bird mimicry, a short anthology of bird passages collected in the Plates. These verbal and visual quotations provide instances of mocking bird technology drawn from and complementing the discussions in the critical essays. Departing from the presumption of merely communicative language, these bird passages are an invitation for readers to develop their own critical and creative lines of inquiry across disparate fields of study.

Bird mimicry is so ubiquitous in pronouncements on poetics, and across so many different cultural traditions, and in so many different historical moments, that it would be impossible to do justice to the full range of examples. While not pretending to be exhaustive by any means, what this collection offers is a constellation of ways to rethink global and comparative studies. As a whole, the volume offers an extended meditation on different specimens of and different senses of bird mimicry: the way birds mimic us, the way we mimic birds, and the way mimicry of any kind involves techniques and technologies of mimicry that extend across as well as beyond languages and species.

"Parrots and Other Birds"

Some of the essays in the volume emphasize bird mimicry in the first of these senses (the way birds mimic us), focusing on the kind of behavior associated with "parrots and other birds," to quote Darwin from *The Descent of Man* (86). The parroting of human speech (which is what Darwin's phrase refers to) is mostly the kind of mocking bird technology we have in mind (as the essays by Holm, Easton, Conway, Meyer, and GoGwilt all foreground). A more extended sense of bird mimicry is always possible, though, even in attempts (like Darwin's) to restrict things to a question of the "language faculty" of humans. So, Frans Snyders's painting *Parrots and Other Birds* extends the linguistic register of parroting to a visual medium of mimicry and camouflage that includes birds not known to mimic human speech. The essays collected here probe the range of mocking bird technologies implied ambiguously by the "parrots and other birds" of both Darwin and Snyders. For our volume, the "other birds" that mimic are mostly starlings and mynahs, and the question of whether these birds can be classified together as a group is both a scientific and a poetic question (to which we will return). So, the volume begins with a pairing of parrot and starling as these famously appear in Daniel Defoe's *Robinson Crusoe* and Laurence Sterne's *Sentimental Journey*, the two examples taken up in the first essay in the volume by Melanie Holm. Other literary examples of birds that mimic human speech appear in Joe Conway's consideration of Poe's raven, Holt V. Meyer's discussion of bird mimicry in Osip Mandel'shtam, and GoGwilt's discussion of the starling in Schubert's song "Impatience" from *Die Schöne Müllerin*. There are numerous further examples one might find, but the economy of this volume's focus on the parrot and the starling provides a template for recognizing the difficulty of disentangling one kind of mocking bird from another. This is a difficulty doubled by the problem of classifying the varieties of bird called "parrot" and by the problem of relating or distinguishing between the "starling" and the "mynah," words sometimes used for the same kind of bird and sometimes designating different birds with distinct cultural connotations.

This difficulty belongs, too, to the difficulty of disentangling bird mimicry of humans and human mimicry of birds. Almost all of our essays turn, in some way or other, to this second sense of bird mimicry: the way humans mimic birds, the way humans represent bird mimicry, or, even more generally, the way humans represent birds. One question that recurs throughout these essays is whether birds can ever escape the cage of human representation. Whether caged or uncaged (and, indeed, whether birds that mimic or birds that don't), this is the question of representation posed by the "parrots and other birds" that Frans Snyders captures on canvas in an aesthetic image of exotic taxonomy and that Charles Darwin tethers to scientific discourse on the human "faculty of language." For both Snyders and Darwin, a newer scientific interest in classifying birds, drawing on much older (indeed ancient) associations, situates the question of human representations of birds at the intersection of a variety of discourses on language, art, and ornithology. As GoGwilt notes in his essay, Darwin's interest in "parrots and other birds" develops a parallel between evolutionary theory and linguistic studies, offering a paradigmatic point of reference for the prominence of comparative philology as a model for scientific discourses of all kinds. All these converge in the matrix of European human sciences, whose formation in the eighteenth and nineteenth century Michel Foucault (in *The Order of Things*) famously describes in terms of an epistemological shift from the age of resemblance to that of representation. Like Snyders's and Darwin's "parrots and other birds," most of the essays in this volume in some way concern the problem of human representation within the framework of the European human sciences that became dominant in the nineteenth century. Each tracks different aspects of this problem of representation at the heart of the European human sciences, following either its formation or its dissolution (and sometimes both simultaneously). This is not to say that the essays themselves are caged within the European framework Foucault characteristically imposes on his critical vision. In examining the problem of representation presented by "parrots and other birds," the colonial coordinates of European discourses are almost always in sight, whether the concern is for birds that mimic humans (e.g., Defoe's parrot) or birds that don't (e.g., Cotton Mather's "Dove," as discussed by Goldberg), and often the concern is for both together, given the ambiguity underlying the mimicry or mockery in human representations of birds.

Homi Bhabha's analysis of colonial mimicry in *The Location of Culture* (1991) provides a key point of reference. Focusing on the colonial mimicry of European discourses, Bhabha notes the "profound and disturbing" effects of such mimicry "on the authority of colonial discourse" (86). The very thing that makes the human sciences hegemonic and universal—their worldwide adoption and adaptation—unsettles the authoritative claim of European origin and superiority. Colonial mimicry threatens mockery: "between mimicry and mockery . . . the reforming, civilizing mission is threatened by the displacing gaze of its disciplinary double" (86). Bhabha's examples emphasize the historical crisis of human representation whereby a fixation on discriminatory colonial classifications makes colonial man simultaneously (and split between) the "object of regulatory power" and the "subject of racial, cultural, [and] national representation" (90). Animal taxonomy and racial taxonomy have long been closely allied, a classificatory confusion (as Bhabha and others point out) that is a foundational feature of the European human sciences. As emerges in discussions of Robinson Crusoe's parrot (Holm and Conway) and the classificatory question of other birds (see Goldberg on Mather's Dove and GoGwilt on Wallace and Darwin), it is this *ambivalence* of mimicry in human representations of birds that makes the European human sciences a disciplinary frame for our investigations into "mocking bird technologies." Many more examples might be adduced to explore how bird mimicry—above all, the parrot—has become a kind of *locus classicus* for the articulation of this colonial ambivalence in postcolonial literature: from *Robinson Crusoe* on up to Joseph Conrad's *Nostromo*, Jean Rhys's *Wide Sargasso Sea*, and Gabriel García Márquez's *One Hundred Years of Solitude* (and *Love in the Time of Cholera*). Bruce Thomas Boehrer's *Parrot Culture* offers an illuminating survey of the parrot trope in such literature (esp. 127ff). Focusing on other bird tropes no less important for postcolonialism, Mustafa's essay examines the neocolonial recapitulations on colonial animal and human classifications that emerge in the mocking figure of birdshit in Yvette Christiansë's *Unconfessed* and the mock-bird's-eye camera view of *Serengeti Shall Not Die*. Extending the postcolonial analytic of colonial mimicry to the full range of techniques used to represent birds—from the verbal and visual forms of early modern print culture up to the technologies of film and digital media—our essays offer models for exploring the multiplicity of linguistic, aesthetic, and scientific questions of mimicry and mockery refracted through the European human sciences.

Studying bird mimicry prismatically through the colonial coordinates of nineteenth-century European humanism, our essays look back and forth historically across a range of theories of representation: classical theories of mimesis; romantic theories of affective, expressivist, and organic forms of representation; and modernist and postmodernist antimimetic theories. Whether birds ever escape the cage of any of these theories of representation remains an ambiguous, open question. The question itself arguably haunts all our examples. And many more might have been chosen. Hitchcock's *The Birds* (1963) surely deserves a mention. Although neither of the two kinds of birds (seagull and crow) used to concoct its narrative of suspense and horror are mockingbirds, the film is premised (both in its narrative and its medium) on problems of bird classification (and representation) characteristic of what we are calling "mocking bird technologies." The *mise-en-scène* of the film involves Tippi Hedren as Melanie Daniels entering a pet shop to pick up a mynah bird. Eager to purchase a bird that can "talk," Melanie's interest in birds shifts—after a seduction scene in which she picks up a man who clearly reciprocates that desire—to a pair of "love birds." Repeating a pattern GoGwilt observes as the "starling" effect lodged in a long history of poetics (across multiple traditions), the "mynah" bird drops out of the picture, replaced by the pair of caged "love birds" (a kind of small parrot) that accompanies Melanie from the beginning to the end of the film. A rather obvious symbol for the love affair between Melanie and Mitch, the caged parrots stand in marked contrast to the flocks of wild birds (gulls and crows; on occasion sparrows and finches, too) that attack humans. The contrast works in multiple ways throughout the film, underscoring perhaps above all the sense in which the film plays on the reciprocal framing of birds and humans: whereas the "love birds" are caught in the cage of human representation, commercial exchange, and human desire, the uncaged birds trap the human actors in a plot of aggressive assault and mass attack.

This contrast between the caged and the uncaged bird—and between the birds that parrot and the birds that don't parrot—reaches far back into histories of poetics from classical Chinese texts (such as Mi Heng's "Rhapsody on a Parrot") and classical Sanskrit texts (such as Bāna's *Princess Kādambarī*) to classical Greek texts (such as Aristotle's *History of Animals*). It is perhaps most pronounced in a distinction Sarah Kay traces in Provençal troubadour poetry between the parrot and the nightingale. In her groundbreaking book *Parrots and Nightingales*, Kay identifies this contrast as foundational in the formation of European lyric poetry. If the parrot comes to signify the mere mimicking of language, the nightingale becomes the model of lyric originality, an ideal of (bird)song to be emulated by European poets especially from Dante on. Kay's argument is particularly revealing in suggesting that the historical emergence of this foundational feature of European lyric is built on a kind of forgetting, suppression, or displacement of the two "ways" woven into the troubadour tradition that shape European lyric form: the "parrots' way" (with its focus on direct quotation) and the "nightingales' way" (with an emphasis on allusion and lyric originality). Kay emphasizes the significance of these contrasting "ways" as models for both creative and critical practice. The poetic techniques of lyric quotation and lyric insertion Kay reexamines in light of this genealogy are inextricably linked to the practices of quotation and anthologization that shape the long histories of European criticism. The technologies of birdsong (whether parrot-talk or nightingale-song) are, in this sense, then, the technologies both of creative, lyric practice and of literary criticism. Given the relevance of her argument for the volume as a whole, we are happy to be able to include a short coda that she has written expressly for the volume.

One practical way Kay's argument illuminates the essays collected here is in drawing attention to the two seemingly different directions in which our essays move. There are essays that foreground what Kay calls the "parrots' way"—those (like Conway's reading of Schreber's talking birds and Poe's raven) that explore the philosophical and poetic questions of talking birds, of mimetic tricks and traps, and of quotation and citation. There are also essays that foreground the "nightingales' way"—those (like Sourgen's analysis of the "volatile unity" of romantic form and Moore's examination of the materiality of lyric form) that explore poetic and scientific questions of birdsong, of romantic expressivist aesthetics, and of lyric form. One might almost divide up the essays into those that examine talking birds and those that examine singing birds. Yet, as Kay's argument points out, it is the interrelation rather than the contrast between the two "ways" that is historically most revealing. Birdtalk and birdsong, though from one perspective entirely different categories of study, emerge from a reading of Kay's argument as two sides of the same coin. Unearthing the dialectic of parrot and nightingale—of quotation and allusion, of lyric creation and critical commentary—in the troubadour tradition, Kay's work suggests a much longer, more interesting, and more complicated comparative historical and linguistic perspective on the European poetics of birdsong. Calling attention to the rich prehistory of human interest in the mimicry of "parrots and other birds," she begins her study by providing an example of the poetic privileging of the lyrical "nightingales' way," which nonetheless illustrates a dependence on the example of the parrot—all the more complicated an example in that the parrot's way is exemplified by one of those "other birds" that mimic, the starling: "Around the middle of the [twelfth] century, the troubadour Marcabru wrote a brace of parodic love songs in which a foolish lover sends a starling—not quite a parrot, but the point is the same—to deliver his message to a tart who then turns him down" (1).

The vexed question of the human representation of birds is what Shari Goldberg traces in her exemplary study of the "avian challenge" of Cotton Mather, the description in *Biblia Americana* of a bird whose precise form eludes critical attempts to reconstruct the mode of its capture in Mather's text. Among other things, this draws attention to the disfiguration of birds that typically accompanies attempts to bring precision to their classification (whether in a philological or in a scientific sense)—one touchstone for this is John James Audubon's method of pinning a dead bird specimen to a framed grid board as a mockup for the depiction of its lifelike counterpart. By contrast to the celebrated success of Audubon's preservation of the live image (even with the death of the actual specimen, or even the whole species, as with Audubon's depiction of Carolina parrots and Passenger pigeons), the disfigured bird can as easily elude classification altogether. This "pigeon effect" (as Goldberg calls it) may seem to take us quite a distance away from the kind of birds that mimic human speech, yet, as Goldberg and others will keep reminding us, the classification of any bird ensnares the human in a fundamental mimetic problem of words. The word for a bird already puts the bird in the place of the word, a mimicry of birds whose logic is as complex and as difficult as the function of language itself. This quite naturally provokes different theoretical and methodological approaches, from Fraser Easton's discussion of the rhetorical capture and release of the "creaturely voice" in Christopher Smart's *Jubilate Agno* to the psychoanalytic and semiotic questions Joe Conway brings to the representation of talking birds in Schreber and Poe. Our essays offer different ways of grasping the foundational logic of this mirror-imaging of bird/human mimicry, and so it is fitting to point to the several possible ways in which this logic lies at the foundations of language and poetics: from the distinction Aristotle seeks to make between the "faculty of language" in humans and (to borrow Conway's citation from Poe) "non-reasoning creatures capable of speech"; to the ancient Chinese association of bird's markings with writing ("When birds' markings replaced knotted cords, writing first emerged" [Liu Hsieh, 10]); to the role parrots and mynah birds play as language teachers in ancient Sanskrit traditions (see, for example, the parrots and mynah birds in Bāna's *Princess Kādambarī*).

It is the chiastic logic linking bird mimicry and the mimicry of birds that also implicates both in a technology of mimesis, representation, or mimicry that might do without (or, more gravely, do away with) humans and birds. The technologies of bird mimicry appear in different forms and with different critical assessments in each essay (whether cast as the merely machinelike imitation of human language [Holm, Conway, Meyer], glimpsed in the technological destruction of actual birds and ecosystems [Goldberg, GoGwilt, Mustafa], or grasped in the form of technology itself [Conway, Moore, Mustafa]). The collection offers two striking examples (among others) in which the bird turns out to be neither mimic nor mimicked but technology itself. Fawzia Mustafa offers a paradigmatic postcolonial instance in her discussion of the chimera of the aircraft as flying zebra, an emblem of the technological apparatus of filming, in which the bird's-eye view is appropriated as the technology of nature conservancy. Isabel A. Moore offers a paradigmatic lyric example in her reading of Dickinson's poem "Split the Lark," in which the lyric subject and object is revealed to be a music box. (This striking example of birdsong technology has its filmic counterpart in Hitchcock's use of Oskar Sala's Mixtur-Trautonium, an electronic musical instrument used to generate most of the bird sound effects in *The Birds.*) A foundational problem of mimesis is embedded in these technological specimens of bird mimicry, and this, too, is evoked by the "mocking bird technologies" of our title.

The deep historical reach of this foundational problem of mimesis is reflected in the pairing of parrot and starling. Presenting a contrast between two species of birds that mimic, the template of this particular difference returns us to the question of whether it is possible to represent a bird at all, a question that extends to the infinitely variable range of questions about how a human representation of a bird might be related to a bird's sense of the human. If the difference between parrot and starling, in one sense (for example, in the contrast between Defoe's parrot and Sterne's starling) offers a merely fortuitous instance of a more general metonymy of "mocking birds" found in different forms throughout the world, the pairing of parrot and starling recalls not only the dialectic of birdsong and bird-mimicry, allusion and quotation, in European lyric form but also the long-established trope of parrot and starling in Sanskrit literary traditions. The pairing is itself a trope from ancient lineages of Sanskrit philology, providing a template for a foundational comparative literary and linguistic problem of mimesis and poesis. In what sense does this pairing echo the phrase "parrots and other birds" in Darwin and Snyders? Already implied in the figurative force of the English verb "to parrot," the metonymy of mocking birds, economically condensed into the pairing of parrot and starling, extends the technologies of linguistic mimicry not only across species of birds—not only across species of animals generally—but also to any species of thing that mimics: machines, for example, or citations (like "parrots and other birds"), or poems, or critical commentary such as this.

Bird: Bird: Not-bird

In the series of seminars from which these essays emerged, a participant sought at one point to characterize all the speakers as compulsively repeating a pattern: beginning with reference to a bird, turning then to consider another (and not always the same species of bird), and concluding with a disavowal (or negation) of the bird in question. The point might be reduced to an almost algorithmic formula:

bird: bird: not-bird

Although the observation seemed intended at the time as a spirited critique, its critical force might in retrospect be taken affirmatively, as a reflection on one of the fundamental insights of the seminar—what collectively, now, as a volume of essays, it might contribute, as an algorithm for measuring mocking bird technologies, to the range of disciplines and areas of critical and cultural studies from which its participants draw their individual approaches. Each essay crosses over the fields of animal studies, poetics, and ecocriticism (to name just a few) in very different ways. Collectively, however, they all address (1) the topic of birds in literature; (2) particular kinds of bird in literature (for the most part, as we have already suggested, parrots, starlings, and other kinds of mocking bird); and (3) the *trope* of birds in literature (which, of course, is not quite the same thing as the bird itself and is, perhaps, the "not-bird" of our algorithm).

Our title captures even as it mimics this movement by classifying the particular birds in our volume as "mocking birds." As noted above, the essays in this volume are not directly concerned with the New World passerine group of birds of the genus *Mimus* or family Mimidae (with the exception of Easton's brief discussion of Christopher Smart's "mockingbird"). Our concern, rather, is more with parrots (of the order Psittaciformes) and starlings or mynahs (of the order Sturnidae)—although we might note that Sibley and Ahlquist classify starlings and mocking birds as "sister tribes in the family Sturnidae" (230). There is a philological precedent for favoring the ambiguous two-word formation of "mocking bird" (see, for example, the *OED*'s citations for its entry on "mocking-bird," especially Bingley's "The Mocking Bird seems to have a singular pleasure in leading other birds astray"). And there is also an authoritative scientific precedent for attending to the wider application of a mockingbird's mocking bird technique. Darwin's breakthrough discovery that the geographical variation in Galápagos finches called into question the "stability of species" originated, in fact, in his observation of Galápagos mockingbird specimens. Although later more famously elaborated on the evidence of the finch specimens he had collected on the voyage of the *Beagle*, Darwin was unable to appreciate the geographical distribution of the Galápagos finches, classifying them initially as different birds because he mistook them for the kinds of birds their plumage mimicked.[2] This classic lesson in the discovery of evolutionary theory already suggests a parallel between scientific and philological interest in "mocking bird technologies." And this, after all, reflects the broader interest in mimicry (and hence mimesis and representation) that led Darwin to draw parallels between the scientific principles of comparative philology and evolutionary theory. We use the term "mocking birds," then, to cover a whole range of birds that mimic, including the spectrum of different kinds of mimicry noted above, from the mimicry of humans (typically the domain of the parrot), to the mimicry of other birds or other species (something mockingbirds, starlings, and mynah birds do), to the mimicry of things (a starling specialty).

We should emphasize that this volume as a whole is very much concerned with turning critical attention to the specific birds in question—whether as a matter of ecology, human-animal relations, cultural history, or biopolitics. It matters, in other words, that the narrower focus of our attention is on "parrots and other birds" that mimic. One of the things that many of the essays return to (albeit in different ways and sometimes with very different kinds of conclusion) is the example set by such "mocking birds" in this looser sense: what they show about human projections onto animals, about human dependence on animals; what they show about language (and the faculty of language in humans as in birds); and what they show (even beyond the question of language) about one of the most fundamental questions of philosophy and poetry, mimesis. In all of these respects, it does indeed matter what kind of actual bird is being referred to, and we want to emphasize here that, although this volume does not follow any one critical path, it does follow the "animal turn" in cultural studies in *this* sense, that it takes real birds seriously. At the same time, it takes seriously the appearance of specific, individual birds in texts alongside their abstractions as species or metaphor, and it also takes seriously the meaning and function of specific birds in their special rhetorical and mimetic functions in literary texts, from the mysterious referent of Cotton Mather's cryptic phrase "*This* was the dove: *this*" to the affective particularity of Robinson Crusoe's "Poll" and Parson Yorick's "That was *my* starling" (Goldberg, Holm).

To take seriously the existence and significance of parrots, starlings, and other kinds of birds that mimic, however, one needs to confront a serious problem of classification. As Paul Carter explains in *Parrot*, European taxonomies of the parrot are themselves highly problematic (arguably more complicated, indeed, than the classification of "mockingbirds"): motivated by an interest in dividing and conquering, they subsume under the sign of the "parrot" the polymorphic character of the bird in question: the two families of cockatoos (Cacatuidae) and parrots (Psittacidae) and "the approximately 184 species of parrot in the Old World and 148 in the New" (23). The value and intrigue of Carter's book is his ability to draw attention to the real parrot "out there" even as it tracks how "parrot" "names a mental creature bred inside the cage of language" (8). If Carter's book focuses on the parrot in "the cage of language," it insists nonetheless on the counterpoint that "*outside* language, whether in their rapidly disappearing jungles or in the eugenicist groves of aviculture—parrots are being hunted and bred to death." "Our culture . . . is systematically pursuing the extinction of its own totem" (8)—this ecological and semiotic argument succinctly articulates a dark thread that runs through all of the essays here, recalling a point Boehrer makes about the mockery inherent in the famous Monty Python "Dead Parrot Sketch" of John Cleese returning a parrot to the pet-shop owner Michael Palin: "Monty Python's dead parrot works purely as a surrealistic exercise in form; symbolizing nothing but itself, it ridicules the very idea of what we might call parrot symbolism" (127). This threatens to turn our algorithmic formula into a sort of grim memento mori, a Monty Python joke to be taken dead seriously: *bird: parrot: dead bird.*

If the polymorphous nature of the parrot as a bird makes a mockery of classification systems, this opens up a series of questions about the parrot's relation to other birds, something Carter's *Parrot* necessarily looks at but also necessarily tends to overlook, given its exclusive focus on parrots. One of his book's early illustrations is from an eighth-century Javanese sculpture depicting a "wishing-tree" in which Carter sees "this-world parrots" paired with "other-worldly bird-women [*kinnari*]" (30). This ancient, non-European depiction of parrots is of special interest because it also shows another pairing Carter overlooks: below the two "this-world parrots" pecking at the top of the "wishing-tree" is another pair, nestled in the tree itself. Are these parrots, starlings, or mynah birds? In many of the essays in the current volume, and as part of the same formula noted above (at least the first part: *bird: bird*), there is a logic that takes us from one kind of bird (the parrot) to another (the starling and/or mynah). This metonymy of birds is particularly relevant for the root problem of mimicry (even before one considers bird-mimicry). Indeed, to simplify a little, it lies as a riddle at the heart of the mirror-imaging of parrot and human Carter underscores when he writes: "We persist in thinking that parrots merely mimic us, when their mimicry is a way of telling us that we are mimics" (8). Carter's point here turns on the scientific observation that "Parrots do not imitate other birds in the wild, and they only talk in captivity" (8). If mynah birds and starlings are like parrots in that they, too, can be trained to talk in captivity (some famous examples may be found in the *Mabinogion*; in Shakespeare's *Henry IV*, Part I; in Laurence Sterne's *A Sentimental Journey* [discussed by Holm]; and in Schubert's song-cycle *Die Schöne Müllerin* [discussed by GoGwilt]), they are unlike parrots in that they do imitate other birds, and even other species and things, in the wild.

There is a repeated motif associating parrots with other kinds of talking birds written into some of the oldest literary traditions—most notably, Sanskrit traditions from which the "śuka" (the green Indian parrot) comes and gives its name to the famous *Śukasaptati* (*Seventy Tales of the Parrot*). So, for example, in Bāna's *Princess Kādambarī* (seventh century), in which a wise parrot becomes the ingenious narrator, there are numerous formulations pairing the parrot ("śuka") with the mynah bird ("sārikā"), as when—before the narrator-parrot appears—we learn that the king's sons are studying the Vedas under the watchful, tutorial supervision of "parrots and mynah birds": "Checked on every word/by caged parrots and mynah birds/who'd mastered all forms of literature" (9). This ancient pairing of parrot and mynah stands in interesting counterpoint to those "parrots and other birds" that appear in the title of Snyder's painting and in Darwin's *Descent of Man*, suggesting another wrinkle in Paul Carter's argument about the polymorphous natural and cultural classification of "parrot" and further complicating his reading of the parrots and other birds in the Javanese sculpture. All these "parrots and other birds" are caught, of course, within the cage of language—multiply so, since we are looking at a figurative play from one of the most ancient of philological traditions, on the one hand, and traditions of pictorial representation and scientific observation, on the other, linked to the emergence of modern critical theory from nineteenth-century European comparative philology.

But the addition of the mynah (or the starling) to the parrot adds something significant in both a scientific and a poetic sense (and it might not be so easy to disentangle the one from the other). The parrot is a bird that "does not imitate other birds in the wild" and is *not*, then, a mocking bird unless caged in human language, while the mynah bird, or starling, *does* imitate other birds—and it does so both caged and uncaged, according to modern theorists and ancient Sanskrit poets alike.

European starlings have featured in recent debates among linguists, cognitive scientists, and animal behaviorists over whether (as Chomsky, Fitch, and Hauser claim) the recognition of recursion (or "recursive syntactic structures") distinguishes human language from other forms of animal communication. To cite the opening from a 2009 article on the topic, "According to a controversial hypothesis, a characteristic unique to human language is recursion. Contradicting this hypothesis, it has been claimed that the starling . . . is able to distinguish acoustic stimuli based on the presence or absence of a center-embedded recursive structure" (van Heijningenab et al., "Simple Rules"). Here, contemporary biolinguistics, catching up with ancient Sanskrit philology and crossing over cognitive science and poetics, singles out the European starling as a bird of special interest for the study of the "language faculty."[3] It is perhaps wise to be skeptical of both sides of the debate about whether starlings prove or disprove the claim for a uniquely human faculty of recognizing complex embedded recursive structures in language. As de Heijningenab et al. argue, reporting their application of the starling studies conducted by Gentner et al. to zebra finches, the tests prove inconclusive. Indeed, they argue, tests of starlings and zebra finches reveal a still more interesting uncertainty: "the same uncertainty remains about human abilities in artificial language learning tasks," leading them to conclude: "At present, there is thus no convincing demonstration of the use of recursive rules in artificial language learning *in any species*" (20541; our emphasis).

There are interesting affinities to be explored between the terms of this ongoing debate within biolinguistics and the terms emerging from the so-called animal turn in cultural studies. So, for example, Jacques Derrida's examination of the difference between reaction and response in the philosophical tradition that deprives "the animal" of language: "All the philosophers we will investigate (from Aristotle to Lacan, and including Descartes, Kant, Heidegger, and Levinas), all of them say the same thing: the animal is deprived of language. Or, more precisely, of response, of a response that could be precisely and rigorously distinguished from reaction; of the right and power to 'respond,' and hence of so many other things that would be proper to man" (*The Animal*, 32). As he reiterates, later, "Kant, Heidegger, Levinas, and Lacan . . . like Descartes, think that in contrast to us humans—a difference that is determined by this fact—the animal neither speaks nor responds, that its capacity to produce signs is foreign to language and limited or fixed by a program" (89). And further, adding a certain critical edge to the scientific claim made by van Heijningenab et al. above, "It is *not just* a matter of asking whether one has the right to refuse the animal such and such a power . . . It *also* means asking whether what calls itself human has the right rigorously to attribute to man, which means therefore to attribute to himself, what he refuses the animal, and whether he can ever possess the *pure, rigorous, indivisible* concept, as such, of that attribution" (135).

Derrida himself does not follow the specific example of the starling. And indeed, despite the affinities linking the biolinguistic debates over recursion in starling song with his deconstruction of the difference between (human) response and (merely animal) reaction, Derrida tends to avoid songbirds (and birdsongs) generally. The exception is itself interesting: his repeated return to Robinson Crusoe's parrot in the lectures collected in *The Beast and the Sovereign*, volume 2, where Poll the parrot is read as a "strange technical prosthesis," an effect of "auto-interpellation," "a sort of living mechanism that [Robinson Crusoe] has produced, that he assembled himself, like a *quasi*-technical or prosthetic apparatus, by training the parrot to speak mechanically so as to send his words and his name back to him, repeating them blindly" (85, 86). Derrida invokes Poll the parrot to repeat the refrain that the animal is deprived of language and is cast as a "first victim of the humanist arrogance that thought it could give itself the right to speech, and therefore the right to the world as such" (260). As Melanie Holm notes in her essay, Derrida seems to miss the nuanced expression of sympathy in the parrot's own words and gestures, displacing any avian expression of human sympathy one might find in Poll's refrain ("poor Robin Cruso," as depicted in the 1719 map of Crusoe's island). Might Derrida's recurrent return to the "prosthetic apparatus" of "poor Robin Cruso" not exhibit *too much* empathy for the poor parrot, revealing in turn perhaps Derrida's evasion of any of the many refrains of songbirds and birdsong? For Derrida, the figure of Crusoe's parrot seems always to be invoked in an act of evasion—"the absolute disaster of this armed word that I shall not even call psittacist, so as not to insult Poll, Robinson Crusoe's parrot (*psittakos*), first victim of the humanist arrogance that thought it could give itself the right to speech, and therefore the right to the world as such" (260).

Derrida's invocation of Poll the parrot as the reductive philosophical refrain (that "the animal means nothing and understands nothing through its cry" [*The Beast and the Sovereign*, 2:220]) might thus be linked to Derrida's evasion of songbirds and birdsongs more generally. As the citation above already suggests, this evasion is connected to the problem of animal classification, a problem especially acute in the case of the parrot (as Paul Carter's book so fully illustrates) and other mocking birds. With Derrida, a particularly interesting touchstone for this problem of animal classification is the work of Deleuze and Guattari in *A Thousand Plateaus* (see *The Beast and the Sovereign*, 1:69). Their celebrated argument about "becoming-animal" (important for a range of debates within and across animal studies) is orchestrated, in certain key ways, around the problematic of birdsong that Derrida avoids.[4] What Deleuze and Guattari call the "becoming-animal" belongs to a broader set of "becomings"—"Exclusive importance should not be attached to becomings-animal"—signaling a shift from organic (and also dialectical as well as evolutionary) models of development to notions of the "molecular," for which the turn away from the romantic trope of "birdsong" serves as example: "The properly musical content of music is plied by becomings-woman, becomings-child, becomings-animal; however, it tends, under all sorts of influences, having to do also with the instruments, to become progressively more molecular in a kind of cosmic lapping through which the inaudible makes itself heard and the imperceptible appears as such: no longer the songbird, but the sound molecule" (248). The Deleuzian "becoming-animal," in this sense, underscores another variation on our algorithm for mocking bird technologies: *bird: birdsong: sound molecule*. To the extent that it is possible to isolate the "sound molecule" from the problems of mimicry and mimesis inherent in representations of birds and birdsong, one might define "mocking bird technologies" as interspecies and paralinguistic, neither restricted to avian behavior nor bound by the presuppositions of human language. Heidegger's notion of "technology," as allied more with the *techne* of poetry (and *poeisis*) than with modern-day conceptions of the "technological," points in this direction. Nonetheless, his reliance on the classical Greek templates of thinking tends to link that *techne* very closely with mimetic problems still bound (perhaps even spellbound) to the logic of human language. To the extent that the "mimetic faculty" might *pre*date the faculty of language, however, as Walter Benjamin suggests in his short essay "On the Mimetic Faculty," the "sound molecule" at the root of "mocking bird technologies" may reveal that area before, above, or alongside language—the paralinguistic zone—to which Benjamin ascribes (in an unattributed quotation) a "reading before all languages": "'To read what was never written.' Such reading is the most ancient: reading before all languages, from the entrails, the stars, or dances" (336). As GoGwilt suggests in his essay, the starling (as bird, in its mimicry of surrounding sounds, and as a trope) inhabits this paralinguistic zone of "reading before all languages," of registering the "sound molecule" of Deleuze and Guattari.

The Starling Trope

The starling trope, in some respects still more so than the parrot trope, belongs at the center of this volume's preoccupations, although it is really the pairing that frames the furthest reach of our topic. As with the parrot, and indeed paired with the parrot, its appearance in literary traditions and in biological reality ("out there"—whether caged or not) poses a range of related but also slightly different questions. The question of the nomenclature—starling or mynah bird—foregrounds a question of cultural difference, of biodiversity, and in the global and comparative perspective of colonial and postcolonial perspectives already posed by the parrot but emerging here in a different way: to begin with, in the decision as to whether to call the bird a "starling" or a "mynah." As with the parrot (and also recalling what Goldberg calls the "pigeon effect"), the word poses a problem of classification. Whereas "parrot" has become the global sign for so wide a variety of birds it signifies a parroting of both bird and word, an exoticism of language and reference from which there is no outside, "starling" and "mynah" both signify a hesitation between bird and word, a split between ordinary and exotic associations—zoologically speaking, they seem to refer to a particular kind of bird ("starling," for example, to the European starling, "common starling," or *Sturnus vulgaris*; "mynah" to any of a variety of different kinds of starlings across southern Asia). Whereas, then, the parrot seems to be the postcolonial bird par excellence, the "starling" or "mynah" reproduces a difference between European and non-European birds and words that challenges all the more our ability to disentangle colonial and postcolonial classifications.

Both parrots and mynah birds have long been the victims of a global trade in exotic birds that threatens the extinction of whole species (as Carter notes in *Parrot* and as is emblematized by the particular example of the Bali mynah or starling discussed in GoGwilt's essay—and, indeed, as suggested by the pet-shop scenes of the Monty Python "ex-parrot" skit and the opening sequence of Hitchcock's *The Birds*). The logic of this human traffic in birds is evidently linked to a wider ecological pattern of imperialism, exploitation, and devastation. Giorgio Agamben provides a compelling formulation for this logic when he extends Foucault's claims about the biopolitics of the modern era (combining those with a reading of Arendt's analysis of totalitarianism) to describe what he calls the "anthropological machine" of modern times that submits "bare life" (*zoe*) to a totalizing administrative control: "The total humanization of the animal coincides with a total animalization of man" (*The Open*, 77). We have a glimpse of this total reduction of the animal to the human, and of the human to the animal, in the "mocking bird technology" of the machine-bird-airplane of *Serengeti Shall Not Die*, which, as Mustafa outlines in her essay and sets in longer historical perspective, reveals the contradictory biopolitical, colonial, and postcolonial logic of big-game wildlife conservation.

Yet even as this *Ndege* (machine-bird-airplane) offers a strikingly apt specimen of the "mocking bird technology" of what Agamben calls the modern "anthropological machine," its chimerical appearance points to another kind of logic—or perhaps we might say, another problem of logic, of logos, of reason, of attempts to rationalize the relation between the human and the animal. Mustafa, indeed, draws not so much on Agamben as on Derrida (as do a number of the essayists, directly or indirectly), specifically Derrida's meditation, in *The Animal That Therefore I Am*, on the elusive figure of "the animal" (for which he invents the neologism "animot")—"Neither a species nor a gender nor an individual, it is an irreducible living multiplicity of mortals, and rather than a double clone or a portmanteau word, a sort of monstrous hybrid, a chimera waiting to be put to death by its Bellerophon" (41). In the posthumously published lectures *The Sovereign and the Beast*, Derrida specifically takes issue with Agamben's distinction between *bios* and *zoe* in a set of critical moves that opens up Agamben's analysis to multiple genealogies of the contemporary impasse of bare life and the state of exception for which Agamben is so illuminating. If there are multiple histories, multiple genealogies of the route that Agamben traces from Greek philosophy to contemporary "Western" biopolitics, Derrida's focus on the "animot"—at least as read in relation to Mustafa's example of the chimerical bird-machine *Ndege*—points a way.

It is neither the way of the nightingale, nor the way of the parrot (nor even "parrots and other birds" that parrot), nor again the way of the bird that is not a bird but its mechanical reproduction. Philologically speaking, it is the way of the starling (mynah, sārikā, jalak, 八哥), marking the multiple passages between philological traditions (European, Sanskrit, Indonesian, Chinese). We should note that here (as in the essays) we touch on only a small cross-section of these philological traditions (so, for example, there are no specimens of the pre-Columbian parrots whose philological significance may be remarked in the trickster role they play in Manuel Angel Asturias).[5] Just as the pairing of parrot and nightingale, in Sarah Kay's account, reopens the genealogy of European lyric form, so, too, we hope that our pairing of parrot and starling—and a sampling of other "mocking bird technologies"—might open for consideration multiple genealogies of bird mimicry throughout history.

Parrots, Starlings, and Other Mocking Birds

Parrots and starlings or mynahs—and "mocking birds" in the loosest of formulations noted above—complicate the area designations of traditional fields of literary and cultural studies. The essays collected here reflect this in a number of productive ways. Two or three of the essays (Mustafa, GoGwilt, and possibly Goldberg) might be classified as embodying a postcolonial perspective, but most of the others follow rather different literary critical affiliations (eighteenth-century studies, lyric studies, romantic studies, Slavic studies, rhetorical studies). This might be surprising in light of the celebrated theoretical leverage given to the term "mimicry" by Homi Bhabha in "Of Mimicry and Man" and throughout *The Location of Culture.* Mimicry as the undoing of authoritative colonial discourse would appear to be doubly, repetitively (or multiply) in question with our focus on parrots, starlings/mynahs, and other mocking birds. As indeed it is—especially when it comes to the trope of the "common starling" that might initially offer the least colonial, or postcolonial, of associations. Yet so fully has the ambivalence of colonial mimicry questioned any monolithic scene of colonizing, or colonized identity, it ought not to be possible any more to study in isolation some European original somehow separate from its extra-European mimic form—or, conversely, to free the indigenous other from the cage of its colonial self-reflection. Robinson Crusoe's parrot (figured at the center of the illustrative map of Crusoe's island) is emblematic for the work postcolonial studies has always already effected at least within English studies. If Robinson Crusoe's parrot provides a fit emblem for a postcolonial philology long at work in defining the colonial coordinates of English, the common European starling offers a complementary image, a kind of negatively equivalent image for the philological commonplace that constitutes the most Eurocentric of European formulations for the study of language. Counterpointing GoGwilt's focus on the starling in Schubert and Sourgen's tracking of starling murmurations throughout European poetic traditions, Holt Meyer offers a fascinating glimpse of the starling both at the center and at the periphery of European poetic forms: in the starling "whistle" of a late Mandel'shtam poem and in the philological trace of the starling/mynah/grackle (*grach*) in Mandel'shtam's reading of the bird formations of Dante's *Divine Comedy.* None of these starlings may be philologically quite the same. The difference, however, is precisely the point. All these birds stage a mimicry of the philological traditions to which they seem attached but may never really quite belong.

Not all of the essays, to be sure, are framed by postcolonial arguments about the hybridity of traditions. Nonetheless, clustered as they all are here under the double sign of parrot and starling, together the essays enact a constellation of postcolonial concerns that have now affected all arguments about mimesis, mimicry, and hybridity—cross-hatched, indeed, with those that concern theories of "affect" (see Holm, Goldberg, Moore), theories of rhetoric and language (Easton, Meyer, Goldberg, Moore), theories of romantic lyric form (GoGwilt, Moore, Sourgen), and last—but by no means least of all—theories influenced by the "animal turn." We are not claiming here that the "starling" tropes that run through all of these essays (explicitly or implicitly) reveal a new turn in the constellation of all these recent directions in cultural studies (we leave the question of attempting a "manifesto" to the Afterword). What we do wish to claim, however, is that the parrot ("poor Robin"'s parrot, and also the parrot of Sanskrit philology) and the starling (both as common starling and as mynah bird, and indeed as the inability to decide between the two), paired together, herald a postphilological, postcolonial, posthuman terrain of literary and cultural studies. Encompassing the most traditional concerns of poetics and the most recent concerns of ecological, animal, posthuman studies, these essays explore the mocking bird technologies by which parrot and starling help us map out the ecosystems of this emerging terrain.

NOTES

1. See Frank J. Sulloway, "Darwin and His Finches," in Donohue's *Darwin's Finches*, 55–97.

2. Ibid., 55ff., esp. 62.

3. The "language faculty" is the phrase used by Chomsky, Fitch, and Hauser in attempting to distinguish what is unique about human language, and it is also the phrase used by Darwin in his meditation on what distinguishes human language from that of "parrots and other birds."

4. David Wills, in "Meditation for the Birds," offers an especially valuable meditation on bird mimicry, along Derridean lines. As Goldberg notes in her chapter, this essay is particularly relevant to the topic of "mocking bird technologies." Another, rather different take on the same philosophical tradition Derrida treats, and also responding to Deleuze and Guattari, may

be found in Bernhard Siegert's "*Parlêtres*: The Cultural Techniques of Anthropological Difference," in his *Cultural Techniques* (53–67).

5. See Carter, *Parrot*, 110ff.

WORKS CITED

Agamben, Giorgio. *L'aperto: l'uomo e l'animale.* Torino: Bollati Boringhieri, 2002.

———. *Homo Sacer: Sovereign Power and Bare Life.* Trans. Daniel Heller-Roazen. Stanford, Calif.: Stanford University Press, 1998.

———. *The Open: Man and Animal.* Trans. Kevin Attell. Stanford, Calif.: Stanford University Press, 2004.

Bāna, *Princess Kādambarī.* Vol. 1. Ed. and trans. David Smith. JJC Foundation. New York: New York University Press, 2009.

Boehrer, Bruce Thomas. *Parrot Culture: Our 2,500-Year-Long Fascination with the World's Most Talkative Bird.* Philadelphia: University of Pennsylvania Press, 2004.

Carter, Paul. *Parrot.* London: Reaktion, 2006.

Darwin, Charles. *The Descent of Man.* 2nd ed. New York: B. Appleton, 1874.

Deleuze, Gilles, and Félix Guattari. *A Thousand Plateaus: Capitalism and Schizophrenia.* Trans. Brian Massumi. Minneapolis: University of Minnesota Press, 1987.

Derrida, Jacques. *The Animal That Therefore I Am.* Trans. David Wills. New York: Fordham University Press, 2008.

———. *The Beast and the Sovereign.* 2 vols. Trans. Geoffrey Bennington. Chicago: University of Chicago Press, 2009, 2010.

Donohue, Kathleen, ed. *Darwin's Finches: Readings in the Evolution of a Scientific Paradigm.* Chicago: University of Chicago Press, 2011.

Hauser, Marc D., Noam Chomsky, and W. Tecumseh Fitch. "The Faculty of Language: What Is It, Who Has It, and How Did It Evolve?" *Science* 298 (2002): 1569–1579.

Kay, Sarah. *Parrots and Nightingales: Troubadour Quotations and the Development of European Poetry.* Philadelphia: University of Pennsylvania Press, 2013.

Liu Hsieh, *The Literary Mind and the Carving of Dragons.* Hong Kong: Chinese University of Hong Kong Press, 1983.

Sibley, Charles G., and Jon E. Ahlquist. "The Relationships of the Starlings (Sturnidae: Sturnini) to the Mockingbirds (Sturnidae: Mimini)." *The Auk* 101, no. 2 (April 1984): 230–243.

Siegert, Bernhard. *Cultural Techniques: Grids, Filters, Doors, and Other Articulations of the Real.* Trans. Geoffrey Winthrop-Young. New York: Fordham University Press, 2015.

van Heijningenab, Caroline A. A., Jos de Visser, Willem Zuidemaac, and Carel ten Catea, "Simple Rules Can Explain Discrimination of Putative Recursive Syntactic Structures by a Songbird Species." *Proceedings of the National Academy of Sciences of the United States of America* 106, no. 48 (December 1, 2009): 20538–20543.

Wills, David. "Meditations for the Birds." In *Demenageries. Thinking (of) Animals After Derrida*, ed. Anne Emanuelle Berger and Marta Segarra, 245–263. Amsterdam: Rodopi, 2011.

CHAPTER 1

"O Friends, There Are No Friends": The Aesthetics of Avian Sympathy in Defoe and Sterne

Melanie D. Holm

to	have	the
say	sympathy	feathered
I	for	friend
can't	another	of
get	one	robin
out	being	robin
again	the	robin
and	starling	crusoe's
again	trope	poll

Both *The Adventures of Robinson Crusoe* (1719), by Daniel Defoe, and *A Sentimental Journey through France and Italy* (1768), by Laurence Sterne, depict unusual expressions of sympathy between men and birds: Defoe's castaway finds himself at the receiving end of a parrot's pity as it cries out, "Poor Robin Crusoe" (166), while sympathy is drawn from Sterne's Parson Yorick when a caged starling exclaims, "I can't get out—I can't get out" (125). The feelings expressed are not unusual in and of themselves; rather, they are all versions of sympathy, an affective experience so lauded in eighteenth-century literature and philosophy as to have been thought identical with a person's moral capacity and posited as the harmonious, unerring law of ideal social relations.[1] The peculiarity inheres in the shift from an exclusively human paradigm to an interspecies dyad that retains human language. Literary expressions of animal sympathy were not uncommon in the eighteenth century, particularly in the equally sentimentalized and ridiculed case of ladies' lapdogs. Nevertheless, the selection of the particular species of speaking animals—parrot and starling—unsettles and reconceives the value and integrity of sympathetic discourse.

By using vocally mimetic birds as agents of sympathy or "fellow-feeling," as it was commonly called in the eighteenth century, Defoe and Sterne raise questions about the nature and necessity of the other "fellow" in the production of feeling that reflect both the crises in sociocultural signification of that century and the uncertainty embedded in the epistemic claims of friendship and other affective relationships. In consideration of the latter question of friendship and epistemology, thinkers such as Jacques Derrida and Giorgio Agamben have drawn attention to the dubious nature of the friend with their examinations of the misattributed apostrophe of Montaigne, "O Friends, there are no Friends" (140), renewing interest in problems of signification, authenticity, and the relationship between affect and mutual knowledge in friendships, particularly as expressed in Aristotle's influential *Nicomachean Ethics*.[2] Aristotle explains that the only true friendship, the friendship of the good, is necessarily rooted in equivalent virtue, "for the friend is another self," he explains, while all other assertions of friendship and social feeling arise from misunderstanding, imposture, and self-interest (9.4.1083). Epistemological surety of the character or interiority of the other calibrates the integrity of affections. Confidently evaluating this criterion, however, proves elusive, leaving Aristotle to argue that true friendship, should it exist, is extremely rare and but slowly cultivated. As a result, anxieties of misunderstanding and deception loom, intimating that where we think we see friends, we have in reality only the "other," inaccessible, alienated, and alienating, that when we call out in sympathy to our friends, we speak only in apostrophe: "O Friends, there are no Friends."

Both Defoe and Sterne exhibit an acute understanding of these epistemic anxieties, crafting moments of what I call avian sympathy, sympathetic interludes between men and birds that transform sympathy's vulnerabilities into productive aesthetic opportunities that value individual human creativity. Each author demonstrates concern for the epistemological and sentimental gaps that may separate individuals where signification is involved, but they approach them with different emphases: deception for Defoe, the integrity of interpretation for Sterne. These authors articulate their complementary concerns for the relationship between knowledge and feeling through the distinct styles of two different species of mimetic birds, presenting two forms of avian sympathy: parrot sympathy and starling sympathy. In the case of parrot sympathy, Defoe uses the parrot's mimicry of human speech to control and vouchsafe signification and meaning in order to achieve otherwise unachievable affective fulfillment. In the case of starling sympathy, Sterne draws from the starling's ability both to mimic and improvise in creating a variety of songs out of one source, in order to posit the pleasure-producing potential of multiplying the meaning-making possibilities of a single sign.[3] With scenes of avian sympathy, these authors inquire into the complex nature of sympathetic feeling not only to destabilize sympathy's status as an authentic and thus absolutely true product of innate moral sense but also to ask what it would mean to think of sympathy as an improvisational, interpretive, and imitative aesthetic act that gives pleasure without certitude.

Sympathy and Signification

Certitude was an uncertain enterprise for the eighteenth century. Throughout the long arc of the early modern period Britain experienced a dramatic destabilization of tacit sociocultural practices that made the relationship between signifier and signified on the levels of language, perception, and social exchange particularly problematic. For many, the concern went further than conscious self-misrepresentation, touching the foundations of human cognition. According to the sensationalist psychology that inaugurated the century, even in cases when signs and senses were not falsified, there could be no guarantee of reliable perception and even less so of uniform understanding. The sensationalist model of John Locke's *An Essay Concerning Human Understanding* (1690) articulates a perceptual process in which all simple ideas are produced by the stimulation of the senses, more complex ideas by reflection on those ideas, and categories and causation by the "association of ideas" (1.2.33). The process, Locke asserts, is vulnerable to the idiosyncrasies of individual sensory faculties, the variable capacity of any individual to reflect upon those ideas, and circumstantial episodes that give rise to "unnatural" associations and superstitions that lead us to misperceive and misunderstand the nature of our world and our experience, "putt[ing] wrong ideas together, and so mak[ing] wrong propositions" (2.11.13).

Locke's concerns for the mechanics of perception and accidental association amplify anxieties about subjectivity, individualism, enthusiasm, and madness, making signification a double gamble for eighteenth-century Britons. The joint unreliability of representation and interpretation poses threats to the integrity of sympathy and therefore unsettles claims of sympathetic sociability in the period because the intersubjective nature of sympathy requires us to make judgments about others based on our perceptions. In the eighteenth century, "sympathy" could signal a variety of experiences, from a form of objective judgment that facilitates emotional response to a sixth, moral sense or affective reflex. This essay will examine the work of Defoe and Sterne as creative anticipations and complements to the thought of Adam Smith in *A Theory of Moral Sentiments* (1759), whose rejection of rectitude and universality and embrace of subjectivity and imagination stands apart from the work of predecessors such as David Hume and the third Earl of Shaftesbury.[4]

Sympathy is largely understood as a communication of sentiments that elicits an emotional response, but complexity inheres because another's feeling cannot be authenticated; it is conceived only through inferences drawn from external signs.[5] In light of this vulnerability, Adam Smith proposes that sympathy must be entirely subjective. Smith maintains that sympathy is an expression of our "natural interest in the fortune of others," but he also separates out the natural from the true, arguing that we develop "fellow-feeling" imaginatively "by changing places in fancy with the sufferer" based on our interpretation of his or her situation as represented to us (1.1.1). The accuracy of sympathetic feelings necessarily stands in uneasy relation to the relative naturalness or artfulness of the object of sympathy and the understanding of the sympathetic observer. In Smith's model, however, accuracy and mutuality are irrelevant because "sympathy does not arise so much from the view of the passion, as from that of the situation which excites it." It is for this reason, he explains, "we sometimes feel for another, a passion of which he himself seems to be altogether incapable; because, when we put ourselves in his case, that passion arises in our breast from the imagination, though it does not in his from the reality" (1.1.1). Indeed, in the case of those who have lost their reason and, as a result, have no understanding of or feelings about their loss, our feelings of sympathy for such an individual become the representative case for all feelings of sympathy—it is but sympathy for ourselves based on how we think we would feel in the case of another.

> The anguish which humanity feels, therefore, at the sight of such an object, cannot be the reflection of any sentiment of the sufferer. The compassion of the spectator must arise altogether from the consideration of what he himself would feel if he was reduced to the same unhappy situation, and, what perhaps is impossible, was at the same time able to regard it with his present reason and judgment. (1.1.1)

For Smith, sympathy neither can be nor is meant to be the reflection of an emotional truth and therefore doesn't require epistemic certainty. As a result, sympathy presents an opportunity for examining projected feelings based on an always speculative assessment of the experience before us. Defoe and Sterne consider such productive possibilities by extending Smith's inaccessible other to the case of nonhuman or differently rational creatures. Their scenes of avian sympathy deconstruct what Smith reads as our natural interest for the fortune of others, so that the fortune or situation of the object of sympathy inspires a mode of thinking about the self that produces not sympathetic feeling for another but pleasurable feelings of creative agency.

The Caged Eighteenth Century

The choice of parrot and starling for these authors' respective inquiries into the nature and naturalness of sympathy is not an arbitrary flight of fancy but a direct destabilization of the common sentimental iconography of birds. In eighteenth-century literature and fine arts, birds were often symbolic figures of sympathy that served as doubles for the subjugated feminine and exotic or wild other. In many cases, birds encode a trope of external beauty without meaningful subjectivity, rationality, or the privileges of masculinity, such as in Edmund Burke's conflation of female and avian beauty in *Observations on the Beautiful and the Sublime.*[6] Similarly, when authors place emphasis on their wild exoticism, birds become symbolic objects of desire and discipline in a hierarchy of hunter and hunted, as in the various avian analogies of Henry Fielding's *Tom Jones.*[7] In colonial propaganda, the parrot and other tropical birds were similarly linked with indigenous populations that were to be contained, feminized, and subdued by the enterprise of empire during "a century which frequently drew facile parallels between Africans and parrots . . . and when the black male bodies most familiar to Europeans were either commodities to own or objects of spectacle" (Nussbaum, *The Limits of the Human*, 195). In domestic relations, caged birds such as the starling were linked with an ornamental femininity that cast women as objects of household display and entertainment for male pleasure. As Diana Donald explains: "The identification of women with caged birds was almost a cliché in eighteenth-century art, based not only on the actual prevalence of this kind of pet-keeping among women, but on the sexual symbolism it contained" (Donald, *Picturing Animals in Britain*, 23). In the influential novel *Clarissa,* Samuel Richardson's libertine villain Lovelace exemplifies this phenomenon, describing birds and beautiful women as lesser creatures equally available for and deserving of domestic captivity, while crediting women with but a greater instinct for self-preservation:

> How, at first, refusing all sustenance, it beats and bruises itself against its wires, till it makes its gay plumage fly about, and overspread its well-secured cage. Now it gets out its head; sticking only at its

> beautiful shoulders; erectedly perched, with meditating eyes, first surveys, and then attempts, its wired canopy. As it gets breath, with renewed rage, it beats and bruises again its pretty head and sides, bites the wires, and pecks at the fingers of its delighted tamer. Till at last, finding its efforts ineffectual, quite tired and breathless, it lays itself down, and pants at the bottom of the cage, seeming to bemoan its cruel fate, and forfeited liberty. And after a few days its struggles to escape still diminishing as it finds it to no purpose to attempt it, its new habitation becomes familiar; and it hops about from perch to perch, resumes its wonted chearfulness, and every day sings a song to amuse itself, and reward its keeper. Now, let me tell thee, that I have known a Bird actually starve itself, and die with grief, at its being caught and caged. But never did I meet with a Woman, who was so silly. (319)

The analogy fancifully extenuates Lovelace's maniacal pursuit of Clarissa Harlowe, who is the sentimental and sympathetic focus of the text. The bird becomes her proxy, eliciting our sympathy in the shadow of such dehumanizing persecution.

As this brisk literary overview of the century suggests, the sympathetic iconography of birds develops in dialogue with the changing sociolingual politics of race and gender in the larger stories of domesticity and colonialism, but it is the anthropomorphic assumptions underlying the sympathetic iconography of birds that Sterne and Defoe question, troubling the status of birds as sympathetic signifiers by taking sympathy itself to task and calling into question the authenticity of the sympathetic feelings they are meant to elicit and direct. The sympathetic relationships between person and bird that they construct unmask the fictional constructs of sympathy in the developing enterprise of eighteenth-century fiction and the sentimental culture it reproduces, demonstrating the agency of imagination that registers as an alternative to objective denotation, a creativity that, like the parroting of the parrot and the song of the starling, seems to feel natural if not strictly true.

Parrot Sympathy and the Displacement of Despair

"Poor Robin Crusoe." Does the parrot actually feel the sympathy he expresses for his human counterpart? The question is certainly a speculative one; the more so since we are asking about the motivations of not only a fictional, inhuman character but a character created and deployed by a writer who was remarkably nimble with, if not also ambivalent about, the deceptive, duplicitous nature of literary fiction and so anxious about the deceptive and duplicitous masks of those around him as to describe his as an "an Age of Plot and Deceit, of Contradiction and Paradox" in which one "can hardly know her Friends from her Enemies" (Defoe, *A Letter*, 10).[8] But for Defoe, the value of Poll's avian sympathy, I argue, lies not in its authenticity but in its effects on Robinson and in its mediating role in the conflict between affectivity and the narratology of spiritual becoming.

To position the nature of Poll's expression properly, it is useful to recall that the parrot is not the first "person" to refer to "poor" Robinson Crusoe on the island. In the first line of his first journal entry, when he describes landing on the island, "poor" is the term Robinson begins with:

> I poor, miserable *Robinson Crusoe*, being shipwrecked during a dreadful storm, in the offing, come on Shore on the dismal unfortunate Island, which I call'd *The Island of Despair*, all the rest of the ship's company being drown'd, and myself almost dead. All the rest of that Day I spent in afflicting myself at the dismal circumstances I was brought to, *viz.* I had neither Food, Horse, Clothes, Weapon, or Place to fly to, and in Despair of any Relief, saw nothing but Death before me, either that I should be devoured by Wild Beasts, murder'd by Savages, or starv'd to Death for Want of Food. At the approach of Night, I slept in a Tree for fear of wild Creatures; but slept soundly though it rained all night. (105)

In his journal, Crusoe adheres to the "mathematical plainness" endorsed by the Royal Society and assumes its quasi-objective posture (Spratt, *History*, 111).[9] Nevertheless, Robinson describes circumstances desperate enough to elicit universal sympathy, and his qualifications of "poor, miserable" indicate that he feels his suffering was a fit object of his own sympathetic feeling. His being "in Despair" on the so-named Island of Despair also suggests that the narrated Robinson felt his situation to be a sympathetic one in the moment, though there was no one there to sympathize with him but himself.

The narrative recollection of the same moment that postdates and revises the journal describes the events with similar pathos but with an increasing emphasis on mental status, and this emphasis interrogates the value of his despair and self-sympathy:

> In a Word, I had nothing about me but a Knife, a Tobacco-pipe, and a little Tobacco in a Box, this was all my Provision, and this threw me into terrible Agonies of Mind, that for a while I run about like a Mad-man; Night coming upon me, I began with a heavy Heart to consider what would be my Lot if there were any ravenous Beasts in that Country, seeing at Night they always come for their prey. (86–87)

a desire for sympathetic sociability, vacillating between a rejection of the sinful or at least pernicious influences of others and a desire for a sympathetic ear. He recounts in his journal:

> I liv'd mighty comfortably, my Mind being entirely composed by resigning to the Will of God, and throwing my self wholly upon the Disposal of his Providence. This made my Life better than sociable, for when I began to regret the want of Conversation, I would ask my self whether thus conversing mutually with my own Thoughts, and, as I hope I may say, with even God himself . . . was not better than the utmost Enjoyment of humane Society in the World. (161)

Though he resolves in favor of isolation from all society but God, he nevertheless marshals his energies to cultivate a partner for humane if not human society: "I did, after some Pains taking, catch a young Parrot, for I knock'd it down with a Stick, and having recover'd it, I brought it home; but it was some Years before I could make him speak: However, at last I taught him to call me by my Name very familiarly" (138). Poll's status alters depending on how Robinson construes his relationship to the bird: shifting from an animal to be captured, to a site of "Painstaking" labor, and, further, to a pleasurable "diversion" such that the bird's speech becomes "an assistance to [his] work" (147). He is then understood as a "sociable creature" and, finally, in Robinson's feudal fantasy, his "favorite" and "the only Person permitted to speak to me" (167, 171). While Robinson's spiritual progress means he is no longer confounded by griefs and joys, Poll's progression from an exotic other to be captured and tamed to privileged confidante suggests that there has also been an aesthetic confounding of the speaking parrot and the "Person permitted to speak."

Identifying Poll as a "person" may serve as a playful allusion to an epistemological gambit of John Locke on the question of what constitutes personality. Following the report of a rational parrot in Brazil who could answer questions and carry on conversations, Locke inquires of speaking birds, "If this parrot . . . had always talked, as we have a prince's word for it this one did, whether, I say, they would have been allowed to be men and not parrots" (2.27.8). The question leads Locke to distinguish between the material form of man and the idea of personhood we usually attribute to human beings, such that "I presume, it is not the idea of a thinking or rational being alone that makes the idea of a man in most people's sense, but of a body, so and so shaped, joined to it," continuing to suggest that personhood or personal identity and human bodies are not necessarily exclusively conjoined:

> This being premised, to find wherein *personal Identity* consists, we must consider what *person* stands for—which, I think, is a thinking, intelligent Being, that has reason and reflection, and can consider as itself, the same thinking thing in different times and places; which it does only by that consciousness which is inseparable from thinking, and, as it seems to me, essential to it. (2.27.9)

Locke concludes that we might consider a "rational" parrot that has a continuous consciousness of experience to be not a man but a person. In Robinson's description of Poll as the "only Person permitted to talk to me," he indicates that Poll attains personhood or a degree of consciousness higher than other animals on the island and that Poll alone is the only one distinguished by speech until the person and man Friday emerges later in the narrative. Indeed, in his reflections, Robinson further emphasizes Poll's singularity by remarking on his difference from other, similarly imitative parrots as well as on the difference of degree of his investments, both laborious and affective, in him: "I had two more parrots, which talked pretty well, and would all call 'Robin Crusoe,' but none like my first; nor, indeed, did I take the pains with any of them that I had done with him" (171). The careful cultivation of Poll's speech and affective performance, however, effects an aesthetic obfuscation of who the "person" is that is permitted to speak and whose consciousness inhabits the bird, resulting in confusion about who it is that calls out to him by name.

> I was wak'd out of my Sleep by a Voice calling me by my Name several times, *Robin*, *Robin*, *Robin Crusoe*, poor *Robin Crusoe*, where are you *Robin Crusoe?* Where are you? Where have you been? I was so dead asleep at first, being fatigu'd . . . that I did not wake thoroughly, but dozing between sleeping and waking, thought I dream'd that some Body spoke to me: But the Voice continu'd to repeat *Robin Crusoe*, *Robin Crusoe*, at last I began to wake more perfectly and was at first dreadfully frightened, and started up in the utmost Consternation: But no sooner were my Eyes open, but I saw my *Poll* sitting on the Top of the Hedge; and immediately knew that it was he that spoke to me; for just in such bemoaning Language I had used to talk to him, and teach him; and he had learn'd it so perfectly, that he would sit upon my Finger, and lay his Bill close to my Face, and cry, *Poor* Robin Crusoe, *Where are you? Where have you been? How come you here?* And such things as I had taught him. (166–167)

Critical approaches to this scene have often dismissed the sympathetic solicitation of Poll's language by reading the initial fear Robinson feels upon hearing another person as a threat to his sovereignty on the island or by focusing on the exchange as empty parroting that prefigures the future uncertainties of language acquisition and cultural assimilation by the native Friday under the dominion logic of a sovereign Crusoe. Many critics share the opinion that *Robinson Crusoe* "is very much a novel in which dominion is the only figure for relations," but this limitation overlooks the sympathetic dimensions in the exchange between Poll and Crusoe and its relationship to his status as an unsympathetic subject of the divine Sovereign (Nash, "Animal Nomenclature," 107).

While Robinson comes to understand Friday as a subject capable of salvation, conversion, and Western civility—and it is for these ends that Robinson instructs him in English—Poll is never nominated as a candidate for conversion. Rather he emerges as a possibility in Robinson's consciousness to fill a desire for sympathy and sociality without giving into despair. For while Robinson's increasing religiosity allows him to speak to God, it is fundamental to the speech act of prayer that it be unidirectional. Faith requires that one never needs a reply,[11] leaving a gap in sociality that Robinson can only ultimately fill with his own creative consciousness and artistic ingenuity.

Surrounded at first only by animals that he is able to domesticate or live with symbiotically, Robinson, as Christine Rees points out, "relates to animals in a way that goes beyond the simply utilitarian" (Rees, *Eighteenth-Century Utopian Fiction*, 85). For example, in the case of a feral cat to whom he offers a biscuit, Robinson is pleased with the pleasure he believes the cat takes in his offering, while in the case of a domesticated goat for whom he cannot find a mate and who outlives her utility, he "could never find it [in] my heart to kill her" and tends to her until she dies of old age. When it comes to the dog that he rescues from the ship, Robinson can find but one problem with the animal traditionally viewed as man's best friend—"I only wanted to have him talk to me, but that would not do." The parrot Poll, however, fulfills the desire for "talk" in a manner that emotionally satisfies and also takes on the value of sympathetic self-reflection.

Poll's speech, "Poor Robin Crusoe," performs a nuanced expression of sympathy in word and gesture that complicates dismissive readings such as Jacques Derrida's claim that "Robinson thought . . . he heard Poll his parrot emit sounds that sounded like words . . . but which were not words, according to him, but noises. And what is lacking in these simple noises . . . is signification. . . . What is lacking to the animal, that of which it is deprived, is signification; the animal means nothing and understands nothing through its cry" (Derrida, *The Beast and the Sovereign*, 308). It may well be the case, as Derrida contends, that "What Robinson thinks of his parrot Poll is pretty much what Descartes, Kant, Heidegger, Lacan, and so very many others, think of all animals incapable of a true and responding speech," but the parrot's truth is not what is at stake in this exchange (278). Rather, it is feeling evoked from the suspension of disbelief in the aesthetic, for if we focus only on the parrot's unknowable capacity to signify inaccessible intents, the significance and affective force of the sympathetic signs are lost.

Setting the Stage for Sympathy

The moment of Robinson's exchange with Poll embeds parrot sympathy in a cathartic teleology. Before he awakes to hear Poll calling his name, Robinson has collapsed, exhausted after a calamitous series of events that he has instructed himself, with some resistance, to read as joys rather than griefs. Having attempted to circumnavigate the island, he ends up on a neighboring islet. Initially impaired from returning by high winds and adverse tides, he espies his former Island of Despair across the implacable sea and reflects on

> how easy it was for the Providence of God to make the most miserable Conditions Mankind would be in *worse*. Now I look'd back upon my desolate solitary Island, as the most pleasant Place in the World, and all the Happiness my Heart could wish for, was to be but there again. I stretched out my Hands to it with eager Wishes. O happy Desart, said I, I shall never see thee more. O miserable Creature, said I, whether am I going. Then I reproach'd my self with my unthankful Temper, and how I had repined at my solitary Condition; and now what would I give to be on Shore there again. Thus we never see the true State of our Condition, till it is illustrated to us by its Contraries; nor know how to value what we enjoy, but by the want of it. (164)

Backsliding toward despair, Robinson corrects himself, and in so doing is forced quite literally to see his captivity on the original island from a different point of view. With this shift in perspective, Robinson also sees himself rewarded when he suppresses his despair and reaches his island, taking a helpful change of winds as he steers back as a sign of providence. He collapses, giving thanks for deliverance, when once again on the island, but this thanksgiving and the providence it signifies become compromised as he discovers himself again to be lost, hot, and discouraged. When safely home, he collapses—this time not out of thanks but out of fatigue—waking only to hear his name repeated to him in "bemoaning tones such as I had taught him" by his parrot: "Poor Robin Crusoe"; "Where have you been?" "How come you here?" The questions speak not to Robinson as a saved soul in Puritan soteriology but as a suffering body and self.

Like many of his adventures on the island, Robinson's salvation ideology iteratively frames this most recent episode as the next step in his conversion, as a test of faith and worthiness, and ultimately as a struggle against giving into despair. Each episode is a signification of spiritual value made before the audience of an all-seeing god. Robinson therefore learns to see and scrutinize himself intersubjectively from the unsympathetic perspective of unmoved and unmovable providence. The performative participation and virtual witnessing of spiritual progress suppress Robinson's suffering and, thereby, deny validation of that suffering as suffering or that he is indeed a sympathetic "poor Robin Crusoe" in the eyes of objective spectators or a reading public.

By teaching Poll to speak "such things" with "such Bemoaning Language," Robinson mimetically reproduces sympathy's vocal and spatial intimacies in a way that is both conventional and meaningful to him as the recipient. Poll lays his bill close to Robinson's face and thereby signifies compassion for "Poor Robin" and concern for his absence and plight. This scene at once suggests an ideal expression and reception of sympathy because the feelings rehearsed in Poll's speech are exactly in sympathy with those Robinson feels for himself but cannot articulate in narratives of sovereignty and salvation. Thus in Poll he creates a surrogate who mirrors back his own feelings and wishes, so that the "person" who sympathizes with Robinson is the only one who can have true knowledge of his situation: Robinson himself.

What transpires between Robinson and Poll only makes explicit what transpires in sympathy's imaginary dyad as described by Smith: "By the imagination we place ourselves in his situation, we conceive ourselves enduring all the same torments, we enter as it were into his body, and become in some measure the same person with him" (I.I.I). The parrot sympathy of *Robinson Crusoe* engages the imaginative agency of the sympathetic spectator, Robinson, so that he who puts himself in the position of a sympathizing other, Poll, creates the sympathy he receives. The intersubjective parody of avian sympathy amplifies the sympathetic fiction, for rather than reading into the parrot's emotions in order to sympathize with Poll, Robinson constructs and projects a self to inhabit Poll, a self that sees him through his own sympathetic eyes. Robinson thereby makes himself both object and subject of a sympathetic pleasure mediated through the parrot's speech in a way that eludes the self-sympathy of despair, for while the words and sentiments are his, the voice is still another's. Thus Poll "parrots" Robinson's sympathizing self-parody. Moreover, even as Robinson embraces these pleasures, he is aware that, although they are illusory and unnatural by the metric of truth, they are nevertheless natural in so far as his instinctive need for sociable sympathy outstrips a desire for the vanities of human society, and the parrot is naturally able to reproduce the words—just as a person might imaginatively reproduce the experience of another.

In a novel of critical reflections, Robinson's narration is surprisingly uncritical of this experience and expresses fond recollection of Poll and his society long after he has left the island. This approbation suggests that the speaker, if not also the author, values this moment and mode of improvisational sympathy as an aesthetic and affective achievement of human art that absorbs and responds like a parrot's revocalizations, that is, not to utter truths but to echo what is desired. Poll emerges as an agent of verisimilar sympathy who parrots back to Robinson the sympathy his struggle against despair rejects while affirming for him that although his soul may be in the process of being saved, his suffering—just or unjust—is not merely a sign of salvation but is the thing itself. After Poll speaks these words to him, Robinson never recollects feeling despair again. And he need not, for he no longer need despair of feeling felt for, even if the feelings are based in fiction.

The Cage of Meaning: Sterne's Starling

Laurence Sterne also presents readers with a captive hero who has a curious sympathetic encounter with a chatty bird, but rather than employing aesthetic parroting as a way to escape the isolation of despair, Sterne's Parson Yorick deductively redefines "captivity" as a condition of freedom by demonstrating a problematic confidence in signs and terms to signify absolutely. The sympathetic pleas of a starling, however, interrupt him at the height of a reverie on rationality and perspicuity, drawing an affective response that unsettles the boundaries of signifiers and signifieds and demonstrating a slippage in signification and play among signs that the starling and his song come to emblematize. When Yorick fears he will be imprisoned in the Bastille for being in France without a passport and thus mistaken as a spy, he draws upon rational deduction to mitigate the horror of captivity, pitting ratiocination against affect: "—And as for the Bastile; the terror is in the word.—Make the most of it you can, said I to myself, the

Bastile is but another word for a tower;—and a tower is but another word for a house you can't get out of.—Mercy on the gouty! for they are in it twice a year.—" (124). Yorick focuses on the signifying power of the word "Bastile" as falsely connotative of imagined terrors, and by reducing the word to its categorical synonyms "tower" and "house," he strips the term of any negative meaning, rendering captivity in the Bastille the equivalent of being home sick with the gout. Continuing in this strain of thinking, he considers the power of the written sign, "Bastile," lamenting the power of the pencil to grant semantic credibility to the mind's hysterical imaginings:

> I walk'd down stairs in no small triumph with the conceit of my reasoning.—Beshrew the somber pencil! said I, vauntingly—for I envy not its powers, which paints the evils of life with so hard and deadly a colouring. The mind sits terrified at the objects she has magnified herself, and blackened: reduce them to their proper size and hue, she overlooks them.—'Tis true, said I, correcting the proposition,—the Bastile is not an evil to be despised;—but strip it of its towers—fill up the fosse,—unbarricade the doors—call it simply a confinement, and suppose 'tis some tyrant of a distemper—and not of a man, which holds you in it,—the evil vanishes, and you bear the other half without complaint. (125)

Sterne's character develops a complex analogy of confinement wherein all confinements of the body are equal and negligible. As a result, the only pernicious forms of confinement are metaphorical and cognitive: imaginative ensnarement in irrational attitudes about objects and experiences that induces fear and other negative emotions and the codification of that ensnaring through the misuse of language effects. Liberty comes through reason, which unpacks the truth of the situation but does so through a linguistic sleight of hand, which ultimately argues that as long as there is liberty of reason, there is no cause for concern over the discomforts of the body. This in turn implies a mind/body division that discredits and denies physical sensation and emotional feeling, including the feelings of sociability and sympathy as well as the spontaneous "delight and surprise," as Joseph Addison puts it, of wit's aesthetic production of coincident, multiple meanings (1).

For Yorick, the autonomous supremacy of the solitary mind falters when an unexpected sensation takes him by surprise. A voice of extreme pathos draws his attention to the feeling of another captive: "I was interrupted in the heyday of this soliloquy, with a voice which I took to be of a child, which complained 'it could not get out.'—I look'd up and down the passage, and seeing neither man, woman, or child, I went out without further attention" (125). Not seeing a person whom the voice signified, he ignores the plea until repetition draws him to their source:

> In my return back through the passage, I heard the same words repeated twice over; and looking up, I saw it was a starling hung in a little cage.—"I can't get out—I can't get out," said the starling.
>
> I stood looking at the bird: and to every person who came through the passage it ran fluttering to the side towards which they approach'd it, with the same lamentation of its captivity.—"I can't get out," said the starling.—God help thee! said I, but I'll let thee out, cost what it will; so I turn'd about the cage to get to the door; it was twisted and double twisted so fast with wire, there was no getting it open without pulling the cage to pieces. The bird flew to the place where I was attempting his deliverance, and thrusting his head through the trellis, press'd his breast against it, as if impatient.—I fear, poor creature! said I, I cannot set thee at liberty.—"No," said the starling—"I can't get out—I can't get out," said the starling. (125)

When Yorick "stood looking at the bird," his rational faculties falter because the source of the sound is so incommensurate with what he had anticipated. But while he is at a loss of what to think, he is at no loss how to feel, at least in the moment.

Like the waking Robinson Crusoe, Yorick doesn't at first know who speaks, and upon discovering the bird and connecting the sound to its source, he overflows with sympathy. Yet in so doing, he reverses the process and turns the source of sound into its own source of meaning. For Yorick, the bird doesn't simply become a self-projection but rather a universal signifier of captivity: "The [image of the] bird in his cage pursued me into my room; I sat down close to my table, and leaning my head upon my hand, I began to figure to myself the miseries of confinement. I was in a right frame for it, and so I gave full scope to my imagination" (127).

A further distinction from Robinson arises in terms of location and population, for while a putative captive, he is not isolated on an island but in a foreign city—nor do the meanings of the starling and his song exist in the exclusivity of a single consciousness. Rather, the scene implies other potential interpretations by the population that comprises "every person who came through the passage" to whom the starling sings its woeful words. Furthermore, the starling doesn't repeat Yorick's words and feelings, though his speech coincides nicely with his situation; rather, the origin, beyond its Anglophone features, is unknown. The intended communication can only be a surmise because of the absence of the original speaker whom the starling mimics and because the starling's interiority, like Poll's, is inaccessible. But the interiority of the starling and the absence of the original speaker become irrelevant: Yorick understands the bird's notes as both "mechanical" and moving, rejecting fixed meaning while inclining toward a prioritization of feeling and playful interpretation. If the parrot talks as a way of imaginatively deploying a fixed meaning and sentiment so that it can return in an ethically permissible form, then starling sympathy, the sympathy elicited by the starling, is an agent of aesthetic multiplication that playfully elicits and inspires a variety of subjective meanings and feelings through the repetition of the same sign.

Rather than a displacement of self-sympathy, the starling invites explicit self-sympathy. One of the more interesting aspects of this curious exchange is how readily the trope of captivity absorbs subjectivity so that the radical otherness of the bird is effaced: Yorick can only imagine the bird's experience in fully human, indeed, in fully personal terms. As Markman Ellis explains, "the immersive animal sympathy imagined by Sterne, when Yorick claims to enter fully into the feelings of the beast, is narrated from the human point of view, and the experience it describes is solely related to human feelings and emotions" (105). On the one hand, the speciesist conditions of possibility for imagination limit Yorick's potential understanding of the starling in human terms, but on the other hand, they allow him to draw together the perceived captivity of the bird, his own captivity, and (in a subsequent passage, "Disguise thyself as thou wilt, still, slavery!") the captivity of slaves as synonymic states in much the same way that he equates the Bastille, a tower, and a house, and imprisonment, the gout, and distemper.

The encounter continues to unleash a series of conventional but nevertheless disruptive affective responses from Yorick, who:

> vow[s] I never had my affections more tenderly awakened; or do I remember an incident in my life, where the dissipated spirits, to which my reason had been a bubble, were so suddenly call'd home. Mechanical as the notes were, yet so true in tune to nature were they chanted, that in one moment they overthrew all my systematic reasonings upon the Bastile; and I heavily walk'd up-stairs, unsaying every word I had said in going down them.
>
> Disguise thyself as thou wilt, still, slavery! said I—still thou art a bitter draught! and though thousands in all ages have been made to drink of thee, thou art no less bitter on that account. (69–70)

While Yorick has the sense that he understands the fiction of the complaining starling, he nevertheless maintains the impression that the bird representatively, reflexively, and necessarily understands him and his impotence, and, what's more, it is only through this network of parodic, sympathetic displacement that Yorick is able to reimagine his own condition and be in sympathy with himself, a selfhood he then distributes across the whole of humanity through what we will see is an ultimately problematic, however deeply felt, analogy for slavery.

The Flight of the Sign

The implication that slavery might wear a "disguise" becomes an ironic feature of Yorick's own initial lack of self-awareness, as his commitment to liberty as an intellectual abstraction translates, on the one hand, to a false confidence in meanings and "systematic reasonings" and, on the other hand, into a liberty of commerce that allows him to sell the bird upon return to England, conflating the starling and the slave more acutely in action than in imaginative abstraction.[12] After his initial description of the starling, Sterne's man of feeling details how, unable to achieve its liberty, he sells it once he returns to England, and then he further relates the bird's career though a variety of owners, from "Lord A" to "Lord E—; and so on—half round the alphabet. From that rank he pass'd into the lower house, and pass'd the hands of as many commoners" (129). The sale of the starling suggestively places Yorick in the subject position of the slave traders rather than the slave and draws questions about his ability to immerse himself sympathetically in the consciousness of a slave or, indeed, to immerse himself in any other consciousness at all, human or animal. The nobility and grandeur of the captive slave analogy fall apart, but in this demystification lies another more potent demystification of signification as a source of credible, absolute knowledge: a playful promiscuity of signification that leaps from one meaning to another, guided by feeling.

The "mechanical" notes the bird sings indicate that Yorick understands that they are not genuine, impromptu articulations of the starling's feelings but a process of mimetic reproduction and improvisation in which he is invited to participate, reproducing the sounds not for communication or ratiocinative definition but for pleasure. In understanding the starling not as a signification but as a site for producing significations, Yorick is able continuously to create meanings for the bird long after he loses sight and sound of him. His creative relationship with its image continues such that he ultimately identifies it as a symbol of himself, claiming, "It is impossible but many of my readers must have heard of him; and if any by mere chance have ever seen him, I beg leave to inform them, that that bird was my bird, or some vile copy set up to represent him. The starling as the crest of arms I have nothing farther to add upon him, but that from that time to this I have borne this poor starling as the crest to my arms" (129). Unable to free the material bird from its captivity, Yorick disembeds its signifying potential from the captivity of perspicuity, ascribing to it a variety of meanings in a variety of contexts, from a sign of his fear of captivity, a sign of human bondage, to a representation of exchange value or pure sign in a commodity market, as well as a sign of self-definition in the most traditional aristocratic terms, a coat of arms that heraldically absorbs the bird as a representation of Yorick's interiority, meaning, and value.

Sterne extends the possibility for creative play with the meaning of the starling beyond Yorick to those others who encounter the bird on the market, for when its sentimental and novelty values wear off for successive buyers, the idea of the bird might invite further combinations of meanings for each self that encounters it, just as it has done with Yorick. The possibility of multifarious meanings suggests an aesthetic play similar to that of the improvisational starling itself, as it repeats, reorders, and reproduces perceived sounds, doing so not to communicate specific content to others but at least in part as a pleasurable act of *poesis* that allows the individual imagination to create rather than determine. Insofar as Yorick does this intuitively, as do the various owners and those who go so far as to make "some vile copy set up to represent" the original bird, his interpretative agency seems less an act of deliberate reason than a natural impulse that operates beyond its purview.

As an idea in the mind, as an impression, the starling has no independent, denotative value, nor can Yorick ever "know" anything about the starling through an act of sympathetic displacement. But it is because of this failure of substantive meaning to obtain in this sympathetic paradigm that its value as aesthetic pleasure can emerge. Sympathy does not achieve real knowledge of others, he shows; it is not an agent of truth but one of fictions, yet fictions that give us the ability better to know, feel, and create pleasures for ourselves.

Conclusion

The concerns shared by these authors suggest that for some eighteenth-century writers in England the idea that we can know another by positioning ourselves imaginatively in their material or corporeal circumstances is pure fiction, indeed a fiction as improbable as a parrot understanding the pathos of Robinson's existential plight and framing it in human language or a caged starling empathetically articulating its plight to a stranger in a foreign country with a language only they understand. Yet if avian sympathy offers a critique of true knowledge, it also illustrates a human capacity and, indeed, natural instinct for creative parody and the pleasures of the imagination. It is perhaps on the level of shared pleasure rather than shared understanding that these episodes of avian sympathies are most evocative, for, in returning to Robinson's return to his island bower, one has the sense that Poll's natural instinct to reproduce "poor Robin Crusoe" alongside Crusoe's other sympathetic interrogatives gives the parrot pleasure concurrent with Robinson's pleasure in hearing and reframing it as sympathy. Similarly, the starling's playful slicing and repetition of "I can't get out" suggests a joy in vocal improvisations that coincides with Yorick's various affective interpolations. Although these moments of avian sympathy may not speak of friendship between man and bird, human and animal (or even human and human, for that matter), they nonetheless create common moments of aesthetic pleasure that are independent of the necessity for, but rather playfully trope, the illusion of understanding amid impermeable epistemic difference. Thus with their curious episodes of avian sympathy, Defoe and Sterne suggest two approaches for sympathetic interplay not merely between man and bird but between writers and readers as well.

NOTES

1. For an historical overview of the sympathy and sensibility in eighteenth-century culture, see G. J. Barker-Benefield, *Sex and Society in Eighteenth-Century Britain* (Chicago: University of Chicago Press, 1992), 149, 108. For more literary analysis of sympathy in the period, see Ildiko Csengei, *Sympathy, Sensibility, and the Literature of Feeling in the Eighteenth Century* (New York: Palgrave Macmillan, 2012); and David Marshall, *The Surprising Effects of Sympathy: Marivaux, Diderot, Rousseau, and Mary Shelley* (New Haven, Conn.: Yale University Press, 1988). For a recent overview discussion of the utopian politics of sympathy, see Michael L. Frazier, *Enlightenment of Sympathy: Justice and the Moral Sentiments in the Eighteenth Century and Today* (Cambridge: Cambridge University Press, 2010).

2. As Agamben has noted, the phrase announces the impossibility of its subject, while for Derrida there are no friends, only others; there is not friendship, only the possibility of hospitality. See Giorgio Agamben, *What Is an*

Apparatus? And Other Essays (Stanford, Calif.: Stanford University Press, 2009), 26–27; Jacques Derrida, *Politics of Friendship* (London: Verso, 1997), 1, 207.

3. For a discussion of the particular mechanics of starling songs among other imitative birds, see David Rothenberg, *Why Birds Sing: A Journey into the Mystery of Bird Song* (New York: Basic Books, 2005), 100–104.

4. For an overview of eighteenth-century theories of sympathy before Smith, see Luigi Turco, "Sympathy and Moral Sense: 1725–1740," *British Journal for the History of Philosophy* 7, no. 1 (1999): 79–101.

5. See commentary in David Hume, *Treatise on Human Nature* (Oxford: Oxford University Press, 2007), 2.1.11.

6. See Edmund Burke, *A Philosophical Enquiry into the Origin of Our Ideas of the Sublime and Beautiful* (Oxford: Oxford University Press, 1998), 3.14.87–88.

7. See Henry Fielding, *The Adventures of Tom Jones, a Foundling* (Middleton, Conn.: Wesleyan University Press, 1975), 1.5, 3.2, 4.3.

8. For a discussion of Defoe's authorial and political relationship with "deception," see Maximillian Novak, *Economics and the Fiction of Daniel Defoe* (Berkeley: University of California Press, 1962); and David Marshall, *The Figure of the Theater: Shaftesbury, Defoe, Adam Smith, and George Eliot* (New York: Columbia University Press, 1986), 79.

9. See the discussion of Defoe's prose style in Ian Watt, *The Rise of the Novel: Studies in Defoe, Richardson, and Fielding* (New York: Penguin, 2001), 113–114; G. A. Starr, "Defoe's Prose Style: 1. The Language of Interpretation," *Modern Philology* 71, no. 3 (1974): 293–294; Ilse Vickers, *Defoe and the New Sciences* (Cambridge: Cambridge University Press, 2006), 99–132.

10. See Michael McKeon, *The Origins of the English Novel, 1600–1740* (Baltimore, Md.: Johns Hopkins University Press, 1987), 319–332; G. A Starr, *Defoe and Spiritual Autobiography* (Princeton, N.J.: Princeton University Press, 1965), 30–50, 105–125.

11. See discussion of Robinson's "apprenticeship in prayer" in Derrida, *The Beast and the Sovereign*, 202–206.

12. See the discussion in Lynn Festa, *Sentimental Figures of Empire in Eighteenth-Century Britain and France* (Baltimore, Md.: Johns Hopkins University Press, 2006), 85–87.

WORKS CITED

Addison, Joseph. *The Spectator* 62 (May 11, 1709). London.

Aristotle, and Richard McKeon. *The Basic Works of Aristotle*. New York: Modern Library, 2001.

Defoe, Daniel. *A Letter to Mr. Bisset*. London: J. Baker, 1709.

———. *Robinson Crusoe*. Peterborough, Ont.: Broadview, 2010.

Derrida, Jacques, and Geoffrey Bennington. *The Beast and the Sovereign*, vol. 2. Chicago: University of Chicago Press, 2009.

Donald, Diana. *Picturing Animals in Britain, 1750–1850*. New Haven, Conn.: Yale University Press, 2007.

Ellis, Markman, "Suffering Things: Lapdogs, Slaves, and Counter-Sensibility." In *The Secret Life of Things: Animals, Objects, and It-Narratives in Eighteenth-Century England*. Cranbury, N.J.: Bucknell University Press, 2007.

Haller, William. *The Rise of Puritanism; or, The Way to the New Jerusalem as Set Forth in Pulpit and Press from Thomas Cartwright to John Lilburne and John Milton, 1570–1643*. New York: Harper, 1957.

Lynch, Kathleen. *Protestant Autobiography in the Seventeenth-Century Anglophone World*. Oxford: Oxford University Press, 2012.

Montaigne, Michel De, and Donald Murdoch Frame. *The Complete Essays of Montaigne*. Stanford, Calif.: Stanford University Press, 1965.

Nash, Richard. "Animal Nomenclature: Facing Other Animals." In *Humans and Other Animals in Eighteenth-Century British Culture*, ed. Frank Palmeri. Burlington, Vt.: Ashgate, 2006.

Novak, Maximillian. *Economics and the Fiction of Daniel Defoe*. Berkeley: University of California Press, 1962.

Nussbaum, Felicity. *The Limits of the Human: Fictions of Anomaly, Race, and Gender in the Long Eighteenth Century*. Cambridge: Cambridge University Press, 2003.

Rees, Christine. *Eighteenth-Century Utopian Fiction*. New York: Routledge, 2014.

Richardson, Samuel, and John J. Richetti. *Clarissa; or, The History of a Young Lady*. Abridged ed. Peterborough, Ont.: Broadview, 2011.

Smith, Adam. *The Theory of Moral Sentiments*. Indianapolis, Ind.: Liberty Fund, 1982.

Spratt, Thomas. *The History of the Royal Society of London*. London, 1959.

Sterne, Laurence. *The Life and Opinions of Tristram Shandy, Gentleman*. In *The Florida Edition of the Works of Laurence Sterne*, vols. 1–2. Gainesville: University Press of Florida, 1978.

Sterne, Laurence, and Katherine Turner. *A Sentimental Journey through France and Italy*. Peterborough, Ont.: Broadview, 2010.

Swain, Kathleen M., *Pilgrim's Progress, Puritan Progress: Discourses and Contexts*. Chicago: University of Chicago Press, 1993.

CHAPTER 2

The Avian Challenge of Cotton Mather's *Biblia Americana*; or, The Pigeon Effect

Shari Goldberg

a
water-
dove

indians
call
them

coecoe-
on-
sogelande

in
my
study

before
me
I

confess
myself
proselyted

this
was
the

dove
this
that

I
have
now

Boston, early January, 1720: Cotton Mather—member of an illustrious Puritan family, famous and accomplished minister, judge, chronicler, naturalist, and exegete—is cold. He is writing an entry in his great tome, the first book of Bible commentary to come out of the colonies, but the work is slow because the ink keeps freezing in its well. He finds it difficult to think in such conditions, but he is warmed by what lies next to his book: a bird, a dove, about which he has a wonderful thesis. In the section of his manuscript on Genesis 8, in which Noah dispatches an avian messenger to locate land, Mather scratches out a description of the bird's form and habits. Then he pauses and uses a pair of tongs to hold the inkstand directly above the coals of his stove, for what he is about to write will require an extra flourish. "This was the Dove," he writes in an italicized script, and then again: "This." He writes a little more and then steps back, cold pen in cold hand, to admire the whole of his sentence: "*This was the Dove: This*, that I have now in my Study before me" (644).[1]

My recreation of Mather's scene of writing cannot dispel the oddity of his sentence: Does he mean, as he seems to say, that he has discovered the dove from the Bible story? Does he mean that the bird before him is literally "the Dove"—Noah's dove? Or does he mean it figuratively, and in that case, how so? Despite these uncertainties, it will seem that I have provided a starting point for interpretation by providing a physical referent for Mather's "*this*." Reading his text outside of my sketch, one wonders *what* and *which* he emphatically indicates by "*this*"; I have suggested that Mather refers to a certain bird, clearly and unremarkably in his line of sight. Yet knowing that a bird was present hardly explains the significance with which it is endowed. In fact, I will argue that the more one delves into the historical conditions through which the bird comes to reside in Mather's study, the more polyvalent the bird appears. By the same token, the more one knows about the events that led to Mather's inscription of "this," the more recalcitrant its meaning becomes: Even as the referent of "*this*" resolves into a single storied bird, the significance of "*this*" branches into many irreconcilable possibilities. "*This*" is the enigma at the center of Mather's sentence, the kernel that generates its uncanniness and resists interpretation.

My essay exposes the stubbornly resistant *this* generated by Mather's avian encounter. Classified as a deictic, *this* tends to point from the text to the scene of its composition. In Louis Marin's formula, "*this* has as its meaning 'the thing which is present'" (16). Hence, to interpret the word, the reader may seek what was present to the author at the time of writing, and having determined the historical conditions, the reader may return to the text believing that she now knows what *this* means. But, I will demonstrate, such neat round trips cannot be made in the case of Mather's text. There is no one set of historical conditions that consolidates the text's meaning; there are rather several interrelated but irreducible contexts, and each adds to what *this* indicates instead of finally explaining it. Even discovering the story of the bird in Mather's study cannot solve the puzzle of his sentences. And so the dove is not an extratextual point of resolution but a figure for irresolution, for the incommensurability of multiple contexts and their resistance to integration.

My meditation on Mather's dove might appear to be an aberration in a collection dedicated to birds that mimic. Yet the tendency of contexts to proliferate resonates with the mimicking bird's ability to multiply the ostensibly singular.[2] Further, Mather's text foregrounds the relay between historical conditions and close exegesis that generated interpretive acts in his day and continues to do so for literary scholarship today. The text thereby speaks to the literary suggestiveness of mocking birds, their ability to destabilize the hierarchy between the original and the supplementary. Of text and context, which constitutes the original and which the trailing echo? Mather's text frustrates an answer, showcasing as well as demanding back-and-forth action, journeys from text to context and back again. Such relay evokes the potential endlessness of mimicry, but it also approximates the habits of a peripatetic bird. Indeed, itinerant birds—such as the dove Mather describes—provide another useful image, in addition to the mocking bird, for considering how the avian complicates the literary. Recalling the circuit of the pigeon, obscure language sends one away from a text, to the historical context, but one cannot help returning to the original site.

I propose the name *pigeon effect* to describe the process of moving to multiple contexts in order to resolve a textual tension that only ends up further complicated. The pigeon effect complements that of the mocking bird, drawing out a related manner in which birds figure the complexities of close reading. Specifically, the pigeon effect raises the question of what it means to contextualize, by challenging the assumption that historical context delivers textual sense. If historical context is not a monolith but a plurality, then discovering it renders the text more, rather than less, open to interpretation. As a result, traveling to the past becomes a means of compounding, rather than reducing, the difficulties of close reading.[3]

Mather's Method

Before undertaking excursions to and from Mather's dove passage, I want to establish his framework for approaching the bird. Mather, too, was an itinerant, in that he traveled between two contexts: that of scripture, which grounded and circumscribed his thinking, and that of his New World surroundings, which required integration into the primary scriptural realm. Toward the end of his career, and especially in his last major work, the *Biblia Americana*, Mather would make a circuit, observing the phenomena of the physical world, such as bird specimens, and figuring their significance in relation to scriptural expressions.[4] Ultimately, the pigeon effect is visible in Mather's writing, but it often seems as if he flits from the physical world into the scriptural realm without accruing complications. For instance, when a set of fossil bones of extrahuman size are discovered near Claverack, New York, Mather turns to an exegete's assertion that "*There were GIANTS on the Earth, in those Dayes*" as evidence that the bones ratify his cosmology: "Could any undoubted Ruines and Remains of those GIANTS, be found under the *Earth*, among the other subterraneous Curiosities, in *our Dayes*, it would be an Illustrious Confirmation of the *Mosaic History*" (582). Confronted with an empirical mystery, Mather seeks the Bible; finding his answer, he concludes that he was right to look there. Mather's tautology importantly protects his belief. He takes something unknown—and therefore potentially threatening to his worldview—and eliminates the threat by demonstrating how well it complements what he already knows.

It may seem that Mather ought to have avoided his trips to the unknown, to save himself the labor of neutralizing his findings. But the excursions were important to Mather: He was committed to studying new specimens in order to identify the scriptural premise that could contain them. Further, he did not need to travel very far for his material. He was surrounded by objects that had yet to be theorized by either European naturalists or exegetes. In 1712, Mather began to provide the Royal Society with letters that broadly engaged scriptural topics but focused on the description of American flora, fauna, and fossils as well as local astronomical, physiological, and anthropological phenomena. While the ostensible object of these letters, known as the "Curiosa Americana," "is frequently theoretical," Otho T. Beall explains, "Mather never lost sight of the fact that he was a colonial writing to the focal point of scientific learning in the empire, nor did he forget that the more peculiarly American the information he imparted, the more valuable would be his contribution" (364). Mather's most compelling communications were therefore about items particular to New England, items that he was uniquely positioned to observe. The Royal Society's response to a bird-themed letter confirms its interest in oddity; according to a published summary, "[Mather's] third Letter relates chiefly to the Birds of that Country . . . He mentions very large wild *Turkies*, some weighing 50 or 60 pound, but the Flesh is very tough and hard. He takes notice of a very large *Eagle* with a great Head, soaring very high, as all of that *Genus* do. As to the Itinerants; he takes notice of vast Flights of *Pigeons*, coming and departing at certain Seasons" ("An Extract," 64). The author seems skeptical about the eagles and pigeons—since all eagles soar very high and all itinerants come and go—but intrigued by the extra-large turkey, which, warranting no comparison, stands out as truly divergent from its European relatives. Thus Mather was professionally invested in confronting what had yet to be explained in scriptural terms.

Mather's commitment to excursion was not without its risks, however, as he met not only with native turkeys but native humans, who often had their own ideas about the birds and the cosmos. As the current editor of the *Biblia Americana* points out (unwanted by publishers three centuries ago, the work is only now seeing print), Mather's title is "momentous" insofar as "the term 'American' (at least until the mid-1750s) signified 'Native American,' rather than 'American of European descent.' The designation 'Native American' . . . pejoratively signified 'Indian' or 'savage,' as the indigenous populations were labeled by European explorers from the moment of first contact" (Smolinski, "Introduction," 75). Smolinski suggests that Mather conceives his project as something like a Native American Bible or a biblical commentary defined by its Native American context. Such a frame wages an alliance with otherness that Mather would seem, culturally, and in this very book, to reject—for if he remained for any length of time within a Native American context, he risked glimpsing an alternative to the truth he regarded as all-encompassing, an alternative that might then suggest his own position as a product of history as much as of divinity. Nonetheless, Mather takes such risks. In the case of the fossils mentioned above, for instance, Mather writes, "Upon the Discovery of this horrible *Giant*, all the Indians, within an hundred Miles of the Place, agreed in a *Tradition*, which they said, they had among them from Father to Son, for some hundreds of Years, concerning him; and that he lived upon the Fish of that River, (usually swallowing *Four Sturgeons* in a Morning for a Breakfast) and that his name was, *Maughkompos*. But there is very little in any *Tradition* of our Salvages, to be rely'd upon" (594). On the one hand, Mather dismisses the pertinence of "Salvage" tradition; on the other, he "salvages" the account when he includes it, and he obliquely refers to such engagements with his title.[5] It is here that the pigeon effect becomes evident in Mather's writing, for his excursion to Native American culture adds a complicating layer to his final assessment. Mather may be working toward a scripturally sound judgment on the bones' origin, but he ends up inscribing a competing explanation that his text cannot synthesize. The bones come to figure a composite of two different interpretive traditions, two different contexts through which they might be understood.[6]

Appropriately, Mather's dove passage provides a more expansive instance of the pigeon effect. The passage culminates in a definitive claim meant to secure the dove in a scriptural context ("*This was the Dove*"), yet to write it Mather interacted with a series of other texts, beings, and objects that remain in its formative shadows. Reading the dove responsibly, I maintain, requires attending to these others and grappling with their tendency to produce an ungainly, unmatching assortment of meanings. In contrast to much literary scholarship, then, I will demonstrate how turning to historical conditions generates more, rather than fewer, interpretive options. I will refuse the common one-way trip from obscure text to explanatory circumstances. Instead, in what follows, I will make several excursions, identifying through each a different context and a different way to make sense of, or to complicate the sense of, Mather's dove. I begin by reading the passage in scriptural terms, then I analyze its relation to Native American experience, and finally I uncover the bird specimen's remarkable history. The series of contexts, I argue, constitutes "the Dove" even as they place increasing pressure on, and ultimately deny, its singularity.

The Scriptural Reading

Mather's dove passage reads as an aberration in the *Biblia Americana*, for while it appears in his commentaries on Genesis 8, the story of Noah's ark, it begins as if it were unrelated to that context. Under the heading "A Discovery made, of the *Dove*, employ'd by our Father *Noah*," Mather provides what appears to be a naturalist's notebook entry on an American avian form:

> We are annually visited in our Countrey of *New England*, at the Beginning of Winter, with a *Water-Dove*, that stays for a little while on our Shore near the Sea, and then goes off again to the Sea, without Returning any more, till the Return of the Year. Tis one of the loveliest birds that fly; and a *Dove* in every thing, except what qualifies it for the Element it is designed for. The *Cock* and *Hen* are paired, as our *Doves* use to be. They are of a deep Azure-Colour; spotted and streaked with White. The Cock has a fine Topple Crown & feathers of a Deep Yellow under his Wings. The Hen is blue, & full of white Spotts all over her Body. Their Noise is a little superiour to that of our Doves on the Land. Their Cooings differ a little from these; about as much as the Voice of a Wild Goose from that of a Tame One. Their soft, yett sprightly Murmurs, are more like the gentle Cadence of Waters, than the melancholy Notes of our common Turtle. They differ from the Doves of the Land, in this; that their Bill is a little flatted: and they have the Membranes between their Toes, wherewith Water-fowl use to be accommodated. When they swim about our shores, they often come upon the Land and they are very frequently seen to have in their Bills, a Small-Branch of our long-lived Cedar, which is green with us all the Winter, & growes on our Beaches as well as on our Mountains. (643–644)

Mather notes the bird's habits, migratory and amatory, as well as its color, shape, and voice. He distinguishes it as a dove, given its form and noises, and more specifically a water-dove, given the "flatted" bill and membranes between the toes, which "qualify it for the Element it is designed for"—the sea. He details the doves' behavior: They often carry a piece of cedar, potentially picked up nearby, potentially harvested from a mountain. Other than the insinuation that the creature was meticulously designed by God, the passage is nearly secular, seemingly unlinked to the promised discovery regarding "our Father *Noah*."

However, after three more sentences, to which I will return, Mather's observations rather suddenly depart from their empirical perspective. He writes that a friend, who has somehow been involved in procuring information on the dove for Mather (more on that, too, to come), advises him to "*be sure, they are one sort of Dove, that found a Shelter, in the Ark; and it is Rational to suppose, that so wise a Pilot as Admiral* Noah *might send forth a Fowl, that could Swim as well as Flie, on the great Occasion we have been appraised of*" (643). Mather's friend's remark returns the reader to the passage's frame, belatedly explaining that the bird has been described not as an American curiosity but because it ought to be considered when one reads the story of Noah. In the first place, if one is interested in all of the animals that traveled on the ark, one ought to note this type of dove; in the second, if one has wondered how the dove Noah sent managed in the gradually abating watery expanse, this dove's form potentially illuminates an otherwise unknown capacity of the genus. Perhaps, Mather's friend suggests, the text refers to a dove like this, one that can "Swim as well as Flie," and not the ordinary land-bound type. This sort of exegesis is right up Mather's alley, and he concludes the passage by approving of the reading. But Mather's agreement, which I referred to above, is startling. He writes:

> Upon the whole, I confess myself proselyted unto the Opinion, that *This was the Dove: This*, that I have now in my Study before me. (644)

This closing sentence is at the heart of Mather's passage. I call it startling for two reasons. First, while the passage had previously read as if observations had been made of several doves, the last line indicates that Mather is speaking of—and more to the point, looking at—a single dove. The single dove, located in Mather's study, might be a live one perched on a windowsill, but his use of the past tense ("This was the Dove") implies that it is a dead specimen. Thus Mather abruptly shifts his orientation from the many live birds that fly, coo, swim, and carry cedar to one that no longer performs any of these functions. The latter dove seems somewhat superfluous to the account, as it is the species' live behaviors that are highlighted in the discussion.

But Mather's concluding sentence is startling, too, for what it claims: that the single dove, the dead one, was "the Dove" of Noah's dispatch. Mather's friend is far more circumspect, writing that Noah "might send forth a Fowl, that could Swim as well as Flie." Mather admits no such condition: He speaks as if the original dove, the one recorded in scripture, has come to him—has literally landed in his study. Even given Mather's belief that the events described in the Bible literally occurred (the pages surrounding the quoted passage are filled with calculations about the size of the ark, the availability of space and feed for various species, and the location of the landing), he claims here that a single dove bridges thousands of years and miles of space. One wonders if Mather reflected on the dove's less than illustrious ending. After participating in events that, he believed, were enormously consequential for the course of human history, the creature had come to rest in an underheated study in a colonial outpost.

From outside of Mather's faith, his statement looks even more suspect: It is easy to smile condescendingly at his italicized claim. Given any sort of ordinary parameters for the life and death of birds, how could it possibly be *the* Dove? Even if Mather is taken simply to be making a more emphatic claim than his friend—to be saying "this was the type of dove"—it seems unlikely that a dove packed into an ark in the Middle East would have made its home on the shores of New England; even if it is possible, Mather forbears to explain the logistics of such a long migration. Like the proposal that he makes, elsewhere, that migrating birds may winter at an as-yet-undiscovered satellite,[7] here he refrains from sustained engagement with nature's physical laws. Thus, regardless of its observed form and habits, the dove's primary context is, for Mather, scriptural. He can refuse other frames—such as the facts of a bird's lifespan and travel capacity—to ally his experience with the time of the Bible. Mather gives priority to the word of his God, rendering its significance all-encompassing.

The Native American Account

Or so it would seem. As I mentioned, Mather's passage includes three more sentences between his observations of the bird and his exegesis on its significance: in the very place, that is, that joins the empirical with the scriptural. Yet rather than providing a seamless transition into the comprehensive world of the Bible, these sentences disruptively introduce a non-Puritan perspective to the account. After noting that the birds often carry pieces of cedar, Mather writes, "They are as delicate a Dish, as we can be Regaled withal; Sweet and Fatt; esteem'd a Royalty among our Natives. Our Indians call them *Coecoe-on-Sogelande*; which is as much to say, *The Messenger at the Great Rain.* This is a little surprizing" (644). The shift from bird behavior to bird consumption renders it an object that sustains the human rather than cohabitates with him. Its edibility is also aligned with the experience of our "Natives": the dish is one with which "we" can be treated "regally," in parallel to the "Native" classification of them as royalty. Mather's parallelism makes it unclear whether the Native Americans indicate the live bird or the meal as royal, and he goes on to introduce another questionable comparison. He provides the Native American name for the bird and then a translation, the messenger at the great rain, which he finds "a little surprizing." Why surprising, and why only a little? The name obviously squares with the role that Mather is about to give the bird, and Mather believes that the Noahic flood was a universal human event, so the information would appear to be affirming rather than surprising. Yet when Mather writes "our Indians," he generally refers to the Abenaki, and the Abenaki do not have a flood myth. So Mather may be surprised that "the Great Rain" is part of the Abenaki nomenclature, but only a little, because he believes that a great rain was part of everyone's experience. He might also be surprised that the Abenaki name seems to confirm an event in which they do not believe, suggesting to him that the flood is indeed universal even if acknowledgment of it is not and that linguistic research might prove informative in this regard.

Mather's glance at the Abenaki introduces a profound ambivalence into the passage, for just as he prepares to make a statement that extends the Puritan worldview, he gives the reader the capacity to question it or at least to stand outside of it. The previous sentences have largely rendered the observing human invisible, appearing to give an unmediated account of the bird's behavior. But as soon as the fact of human interaction becomes unavoidable—when the bird becomes an object to eat—Mather refers to a diverse community of eaters, the Puritan "we" apparently appreciating the bird as much as the Native others. The fact of this community is extended as Mather provides the Abenaki name, indicating that they recognize and have classified what the colonists are only now noticing. Mather then shifts to offer a series of judgments that further place the Abenaki in conversation with the Puritans. First he judges in relation to the Abenaki name—"this is a little surprizing"; then his friend judges in relation to the bird's form—"be sure, they are one sort of Dove, that found a Shelter, in the Ark"; and finally Mather judges in relation to scripture—"This was the Dove." The Abenaki perspective introduces the need for judgment, producing a triangular arrangement whereby the Abenaki, the friend, and Mather surround the bird and offer assessments of it. Mather's final claim is undoubtedly meant to be the last word on the subject, but the passage suggests that Mather only arrives at it after engagement with others—and, more pointedly, with otherness. And since he does not outright reject the otherness, it is as if the exposure to difference somehow makes his one-sided claim, even as it would seem to prohibit it. Resisting coherence, Mather's passage arrives at exclusion via inclusion, arrives at sameness via otherness.

The History of the Bird

The complexity of the water-dove passage may be traced in historical as well as rhetorical terms. The set of events that led Mather to write the passage is, to use his phrase, "a little surprizing," in that it demonstrates that encounters with difference and confrontations with otherness underpin the passage even more extensively than my close reading has heretofore suggested. Mather's passage turns out to be largely written by someone else, Mather's bird turns out to be not only dead but dismembered, and Mather's account of the Abenaki names turns out to be at best confused and at worst deeply wrong. I will turn to each of these complications in turn, illustrating how they add layers of history to Mather's passage and complicate our ability to read his final sentence accurately—that is, to know what, exactly, is meant when Mather refers to "*this*" dove.

A Copied Letter

As I noted earlier, Mather's passage on the water-dove stands out from the others surrounding it in the *Biblia Americana.* This divergence may be the result of Mather's having copied rather than originated almost all of its sentences. The water-dove was procured for Mather by one John Winthrop, of New London, Connecticut, who accompanied his shipment with a letter providing almost all of the details that appear in the *Biblia Americana.* In historical accounts, this John Winthrop is distinguished from the brethren bearing his name by the designation "F.R.S.": Friend of the Royal Society. So Winthrop was qualified to observe the dove. But Mather's wholesale adoption of Winthrop's report necessarily muddies some facts; for instance, Mather begins by situating the dove "in our Countrey of New England," but surely it would be more accurate to locate it in Connecticut, for if the bird also frequented Boston, Mather would not have needed Winthrop's material or information. It seems possible that Mather intentionally obscures the dove's origin, a hypothesis that is supported by other subtle adjustments of Winthrop's prose. Comparing the two documents, Mather's changes appear so minor as merely to erase the idiosyncrasies of his source text. Here are the sentences and phrases of Winthrop's that Mather copies or adapts, in order of appearance:

> The Hen is Blue and full of Small White Spotts all over the Body.
>
> But the Cock has a fine Topple Crowne . . . [a]nd has one of the Gravest Coulers of the Rainbow under its Wings.
>
> Their Meat is tender Sweet and good and very fatt; and I have Eaten of them Severall times: They are Reckned a Royalty among the Natives.
>
> They make a more Superior Noise then our Land Doves doe and their is as much Difference between the Coeings of these and our common Pigeons, as there is between the Voice of a Wilde Goose and one of our Ordinary Game ones.
>
> Their Soft yet Sprightly Murmurings seem more like the Gentle Cadence of Waters then the Melancholy Notes of the Turtle.
>
> When they Swim about the Shores they often come up on the Dry Land and are frequently Seen with Small Branches of the Long-Lifed Cedar in their Bills: the Cedar is green all Winter and grows here on the Tops of the Highest Mountains as well as in the Lowest Valleys or in Wett places as Swamps and often on the very Beaches.
>
> Whether they Eate the Buds thereof, or it be the Emblem of which wood the Arke was made of which they pull off, or pick up under those never Rotting Trees to sporte themselves withall, I Leave you to Determine.
>
> Be Sure they are one Sorte of Dove which Anciently was Sheltered in that Vast Floate; and its very Rationall to Suppose so skillful a Mariner and Wise Pilott, as Admirall Noah was, would send forth a Fowle that could Swim as well as Flye. (Winthrop)

I am not interested in indicting Mather as a plagiarist; given his moment and his engagement with volumes and volumes of texts, he ought not to be expected to be scrupulous about citing his sources. In fact, Mather may actually have indicated his source, if not cited it. When he italicizes his friend's concluding remark, it seems as if it is a direct quotation, which implies that the rest of the passage is not. Yet Mather introduces the remark by writing "And my Friend from whom I have this Communication, together with the agreeable Sight of the Bird, concludes with this Remark." "This Communication" may reasonably be read to include all of the foregoing, with the verb "concludes" indicating the end of the communicated report. Nonetheless, it is worth noting that, in a letter asking Winthrop for a description of "your water-dove," Mather writes that he wishes for "my account of it [to be] authentic . . . and [to] have your name (which I snatch at all occasions to do honor to) incorporated into it" (Silverman, *Selected Letters*, 300). Mather certainly seems to have overpromised his friend—the more so as when he writes to thank Winthrop, he assures him that he has made in the *Biblia Americana* "an honorable mention and character of the generous hand which like another Noah, sends [the water-dove] forth unto us" (Silverman, *Selected Letters*, 301). "My friend" hardly seems such laudatory mention.

In any case, the copied sentences reveal that Mather was editing and collating as much as writing: He was engaging with another voice, then, and another person, even in the naturalist-sounding sentences that minimize the presence of the observing human. While it seemed, in my analysis above, that the passage begins with the dove and then shifts to include an eating human and then different types of eating and judging humans, the passage in reality records a human conversation from the start. The relationship between Mather and Winthrop, as collaborators, is as fundamental to the passage as the dove itself. Hence, the passage may be considered a composite production issuing from at least three languages: that of Winthrop, that of the Abenaki, and that of Mather. It is a text generated out of differences even as it purports to subsume them.

Mather's rearrangement of Winthrop's sentiments is also worth noting, as it reveals that he organized the passage such that it proceeds from the dove to the humans. As the list above demonstrates, Winthrop refers to eating the doves, and to the Native American appreciation for such meals, in between descriptions of their form and their voice. By contrast, Mather does not refer to the birds as food or as part of a cultural system until he has given a full account of their appearance and behavior. Thus it appears that Mather understood, as I suggested earlier, the engagement with the Native American language as preparatory for his judgment on the bird's place in his own scripture. Centuries before Wallace Stevens, he was crafting a text about different ways of looking at a dove.

A Dove in Parts

The few sentences from Winthrop's letter that I did not transcribe above explain how the dove came to be delivered to Mather's Boston study. Winthrop writes that a man named Richard captured a male dove, which, it seems, was to accompany a female dove already procured for Mather. But "a Mischievous Catt" intervened: It got into Winthrop's study and tore the female to pieces. Winthrop writes that the cat was "Like to have Spoilt [the male] alsoe. It has pulled the Neck from the Body and much Darted it; but it may be fastened together so as you may see the Figure of it." This detail means that the dove to which Mather attributes such an illustrious past must have been a grisly sight. Even if the "dreadful cold" of the season (which I earlier evoked; Mather mentions it when he thanks Winthrop for the package [Silverman, *Selected Letters*, 301]) arrested severe decomposition, Mather had before him a punctured dove in two parts, head and body, which he may or may not have fastened together.

This violent history underwrites Mather's text as much as his scriptural compass, for the dove only arrives in Mather's study after Richard has captured and killed it and the cat has decapitated it. In order for Mather to encounter the embodiment of God's word, that is, the body must be subject to rather horrific mutilation. Put even more strongly, the mischievous cat makes Mather's passage possible as much as Noah does: Both craft a part of the story, a chapter of the bird's life and death. And even as Mather followed these chapters with his own, he too was suffering materially. "'Tis dreadful cold," he writes to Winthrop. "My ink-glass in my standish is froze and split, in my very stove" (Silverman, *Selected Letters*, 301). The mutilated bird, the freezing man: These are the conditions under which Mather declares his discovery. These conditions are, moreover, particularly American conditions: In addition to being locatable in the colonies, they refer to the violent deaths and the bodily suffering that marked the experience of early eighteenth-century America. The decapitated bird may even register, obliquely, the scalping, the violation of the head after death, with which Native Americans responded to colonial audacity. Headless yet inscribed in God's book, the water-dove metonymizes the alternation between physical violence and pious practice that characterized Mather's generation. In the textual web I am tracing, Mather's biblical commentary records these material realities. This was the dove, we might amend Mather's words, that Noah sent to find land and peace, which then suffered the American fate of losing its head to a sharp, pointed weapon.

A Vexed Message

There is also violence, or at least misunderstanding, to be traced in Mather's explicit reference to the Native Americans, his claim that they call the dove "*Coecoe-on-Sogelande*," which he translates as "Messenger at the Great Rain." Mather's source for this name is difficult to identify. Winthrop refers to the dove as "the Sogelander" in his accompanying letter, suggesting that he has previously introduced Mather to the idea of a dove called something like Sogelande or Sogelander. Such a name would seem to derive from the Abenaki word *soglon*, which does indeed, as Mather implies, mean rain. Mather's prefix is less legible; it is possible that Winthrop's earlier communication about the dove included the longer name. Yet it doesn't seem likely that either man possessed a completely accurate transliteration. "Coecoe," or "Coecoe-on," does not, as Mather implies, mean "messenger." The Abenaki word for messenger is *kiwagimok*, which sounds something like—but not very like—"coecoe." Winthrop describes the noises the doves make as "coeings," so it is possible that the Abenaki used onomatopoeia to refer to the bird, as they do with the word for owl, *kokokhas*.[8]

Even this rather generous reconstruction is incomplete, however, as the Abenaki name for pigeon is *pelaz*; as far as I can tell, there is no separate word for *dove*. The Abenaki bird name that sounds closest to *sogelande* is *soglonihasis*, which means swallow. Indeed, the bird that Winthrop and then Mather describe has features in common with the swallows found in Connecticut, especially the deep azure color and the white spots and streaks. So perhaps one culture's pigeon is another's swallow. In Mather's commentary on Jeremiah 8:7, he addresses a related discrepancy, arguing that the King James translation mixed up the Hebrew words for *swallow* and *crane*.[9] It may have been difficult for Mather to be sure that he held either a pigeon or a swallow, given the condition of the body he had. It might also be the case that Winthrop gave the name "Sogelander" to the bird in confirmation of his thesis that Noah would have sheltered and made use of it.

To some degree, then, Winthrop and Mather appropriate the Abenaki language to their own ends, and the passage thereby commemorates such acts of cultural violence. At the same time, since the parameters of their appropriation remain unclear, the word *sogelande* may also be said to mark the slippage of language, and the incommensurability of conceptual categories, between the colonists and the Native Americans. Mather acknowledged such encounters when he advertised the *Biblia Americana* to potential British publishers. Referring to texts that were published and read in Europe and only later made their way to New England, Mather urges, "The Writers whom you made much of, while you had them *at Home* with you in a more *separate Condition*, certainly, will not lose your Favour, for having *Travelled Abroad*, and now *Returned Home* in Company; tho' with their *Habit* and *Language* having something of an *American Change* upon it" (34). Mather suggests that when he reads and interprets the European texts, they gain an "American Change," becoming different and worth reconsidering. If we recall Smolinski's note that "American" meant "Native American" to Mather's audience, then Mather suggests that what happens to the commentaries in his study is that they become subject to, and changed by, Native American influences.

Mather steps into unlikely territory when he proposes that a text reads differently in America and in Europe. One reason that his claim to have found Noah's dove is so striking is that it appears to disregard external context completely: Unaffected by the years or the miles, unaltered by any relationship with a nonbelieving set of peoples, the dove's meaning is statically sealed in a scriptural compact. The dove's Americanness seems only incidental; Mather is certainly proud to have found the bird in his region, but the power of God is for him such that the dove might have been found anywhere. Yet if, when Mather reads a book, he comes to write about it in a particularly American way, then the dove's place of habitation also endows it with "an American Change." It becomes a dove altered and thus defined by its immediate surroundings. In this sense, the history of the dove that I have been tracing, which Mather may have wished to exclude, actually constitutes his text. The dove is produced by virtue of its American context: by letters sent from Connecticut to Massachusetts, and a local mischievous cat, and exchange or confusion between English and Abenaki. In pieces, it derives from Winthrop and Mather, and from the cat and the Abenaki, and its split condition is as pertinent as the holy one Mather meant to inscribe.

Upon the Whole

"Upon the whole," Mather writes, "I confess myself proselyted unto the Opinion, that *This was the Dove: This*, that I have now in my Study before me." But, as I have explained, the bird was not whole at all, and the whole to which Mather refers—all of the information from all of the sources that generate his final surmise—is in no way smoothly integrated. Further, Mather's deictic "this," capitalized and italicized, means to point to one object, to *the* bird: "*This*, that I have now in my Study before me." Yet "this" refers to a collection of differences. Mather may be looking at one physical form, but when he turns to inscribe it on paper, when he renders his emphatic "this," the avian breaks into several directions at once. "This" means: the dove caught by Richard, the dove mutilated by the cat, the dove named by the Abenaki and then by Winthrop and then by Mather, the dove that circulated in the Native American environment and was put to use in the Christian cosmology, the dove that Mather could use to seduce the naturalist establishment of the British empire, the dove that was neither Christian nor American but whose textual history marks the crossings of these identities in New England in 1720. "This" means: the dove of several letters, of several vocabularies and several books, the dove that many had spoken and written of before Mather undertook its commemoration. "This," a grammatical term used for identification, ultimately disidentifies its object, making it impossible to gather its separate parts, to fasten its components together without leaving a noticeable and storied scar.

It is fitting, then, that Mather's specimen remains unidentified. It would have been more than "a little surprising" had the bird's form been preserved. But Mather's object seems to have disappeared entirely. No scholar has identified what animal, exactly, Mather had before him—what species of bird it was that he called "water-dove." Such doves may be extinct, or, as I suggested above, they may be swallows. In any case, they are unknowns; they cannot be called upon to help make sense of Mather's uncanny passage.

Postscript: This

I have tried to articulate a list of all that must be considered the referent to Mather's "*this*." The list form approximates my own experience of reading and researching, of coming back to the emphatic word with more and more that it might mean, more and more that could not be indicated with a single gesture or mindset.

In fact, I repeatedly found myself stymied by the word "this" in the course of my research. Again and again, in a variety of texts, I had the impression that if I could only be certain of what "this" meant, I would be able to draw my analysis to a close. But "this" *always* turned out to point to multiple referents at once. Here are four instances, the first two of which I briefly mentioned above, that I found most promising and most stymieing:

1. Mather's raised eyebrow with regard to the Abenaki name of the water-dove, "This is a little surprizing," seems to refer to his translation ("The Messenger at the Great Rain") but likely responds also to what it implies about world religions. As Jan Stievermann explains, for Mather, "all pagan religious beliefs and practices in the world . . . derive from Noah's faith. In other words, they are more or less depraved corruptions of the original, monotheistic religion established by the patriarch after the flood" ("Genealogy," 536). So it would not be surprising to him that Abenaki culture preserves a remnant of something called "the Great Rain." But since the Abenaki do not believe, as Mather did, that their history could be encompassed by the Mosaic stories, it may indeed have been surprising that the two cultures happen to coincide in associating

the dove with a flood. What surprised Mather more: that the Abenaki did not recognize his mythology or that they did?

2. Mather's nod toward the end of his passage to "my Friend from whom I have this Communication" is also unclear. "This Communication" may refer to the entirety of the foregoing passage or to the specific information about the Abenaki's view of the bird. "This Communication" may even refer to the dove itself, which was transported from Winthrop's study to Mather's, as per the now obsolete definition of *communicate* as "to give or offer (a material object)" (*OED*). How much credit is Mather obliquely giving his friend here? The remainder of the sentence is odd as well; Mather writes, "And my Friend from whom I have this Communication, together with the agreeable Sight of the Bird, concludes with this Remark, *That be sure, they are one sort of Dove.*" "Together with" may mean something like "based on"; otherwise, Mather would seem to be saying that his own visual evaluation of the bird is agreeable to the friend's conclusion. Given the strange appositive, it sounds as if Mather's own sighting interrupts the presentation of his friend's judgment and that both men conclude the one idea. Perhaps, then, it is important that Mather *has* "this Communication"; it may be that the act of communication articulated is the transfer of the idea (that the water-dove belongs to the Bible story) from Winthrop to Mather.
3. In Winthrop's letter to Mather accompanying the dove, he explains the damages wrought to the specimen by writing that the cat "has torne the Hen pigeon in pieces and Like to have Spoilt *this* alsoe. It has pulled the Neck from the Body and much Darted it; but it may be fastened together so as you may see the Figure of it" (emphasis mine). In his influential biography, Kenneth Silverman summarizes the letter by writing that "[Winthrop] told Mather, he was going to send a water-dove itself but a cat got into his study and tore it to pieces" (*Life and Times*, 247). In surmising that no dove was sent, Silverman seems to have missed Winthrop's "this," which is followed by no noun but which surely refers to the mutilated male creature he had enclosed. Reiner Smolinski, in a footnote on Mather's water-dove passage, figures that an actual dove must have reached Mather's study for him to claim that he has it there before him, but Smolinski suggests, following Silverman, that it must have been another, presumably intact one sent at another time (Mather, *Biblia Americana*, 644n44). I seem to be the first scholar to claim that Winthrop's letter narrating the gore was paired

with a dove and that, consequentially, the dove in Mather's study was the much-damaged one. The note "wth. a box," which appears on the envelope of Winthrop's letter, provides evidence for my assertion.

It is unlikely that some other object was enclosed in the box and that another box, with an intact bird, separately reached Mather swiftly enough for him to reply with a thank-you note twelve days after Winthrop composed his letter. Yet for Silverman, Winthrop's "this" was overlooked or referred to nothing at all, which caused Smolinski to think that "this" dove that Mather had in his study before him was new, unrelated to the letter. I assert that the additional, intact dove that Smolinski infers from Mather's "this" never existed; moreover, its invention unintentionally whitewashes the bloody conditions that give rise to the text.

4. In Smolinski's aforementioned footnote, I misunderstood the word "this" to indicate that an earlier draft existed of Mather's water-dove passage. Smolinski writes, "Mather's scientific description of the American passenger pigeon (*Extopistes migratorius*)—now extinct—appears in his *Christian Philosopher* ('Essay 30,' pp. 204–08). An earlier draft version of this essay intended for the *Philosophical Transactions of the Royal Society* is included in Mather's collection of 'Curiosa Americana' (1713–1717), in the MF Edition of the Cotton Mather Papers (Reel 5, item I #5)" (Mather, *Biblia Americana*, 644n44). I took the "this" in the second sentence to refer to the passage supplemented by the footnote, rather than the essay mentioned in the sentence prior. I wrote to Smolinski via e-mail and asked where I could find the draft. He replied with the following (it is important to note that this text, including all emphasis and parentheticals, is directly transcribed from Smolinski's message):

> RE my statement in note 44: "An earlier draft version of THIS [emphasis here added] essay intended for the Philosophical Transactions . . . is included in Mather's collection of 'Curiosa Americana' (1714–1717), in the MF Edition of the Cotton Mather Papers (Reel 5, item 1 # 5)." The pronoun THIS refers to the description of the PASSENGER PIGEON (in Solberg's edition of Christian Philosopher (204–08), NOT to "The Water Dove." The "earlier draft" of this description is on Reel 5, item IX [[NOT item I (mea culpa)]] #5.

I envied Smolinski his certainty about what "the pronoun THIS" indicates, although his acknowledgment that "item I" could mean

> "item IX" fit the pattern of my research. Rather than lead me to another source that would help explicate Mather's "This was the Dove," Smolinski's response turned me back to the uncertainty in the text I had already read so many times.

My "this" collection was generated in the course of research that, at first, was oriented toward tracing Mather's bird in order to grapple with the history of birds in New England, especially their extinctions. But my pigeon researches generated a pigeon effect. It became impossible to learn about Mather's bird without confronting the fact that I still did not know—that I was in fact losing certainty with regard to—how to situate a single disappeared animal; I could hardly generalize further. I was a little surprised to find myself, like Mather, determining that the facts of a bird's life ultimately mattered not because they led me to understand its past but because they involved me in an extraordinary interpretive undertaking.[10] I propose this as the legacy of Mather's dove, or of his dove passage: an imperative to recognize that the challenge of reading birds is always a challenge of reading words.

NOTES

1. My recreation is drawn from a letter Mather wrote on January 11, 1720, printed in Silverman, *Selected Letters*, 301.

2. David Wills summarizes the questions raised by bird mimicry in his essay on recorded bird song: "Are the birds making music, or simply mimicking, parroting or aping themselves? Is their song a call and response, a chant, or simply a repetition? Is it a theme and variation, indeed an improvisation, or rather a mechanical repetition, or indeed reproduction?" ("Meditations for the Birds," 245).

3. Why pigeon effect and not dove effect, following Mather's passage? I use the former term in part to refer to the habits of actual birds, of the family Columbidae, as separate from the fraught figure of the dove at work in Mather's text—even though I will largely unravel this distinction by the end of my essay. Another reason for emphasizing pigeons over doves is to pay tribute to the passenger pigeon, about which Mather also wrote, which was plentiful in North America in his day but rendered extinct because of excessive hunting practices by the early twentieth century. See Christopher Cokinos, *Hope Is the Thing with Feathers: A Personal Chronicle of Vanished Birds* (New York: Tarcher, 2009). It is also worth noting that an itinerant-based effect is appropriate for the colonial moment and for Mather in particular, whose *Biblia Americana* is touted as "the first comprehensive biblical commentary composed in North America" (Stievermann, "Introduction"), even as the texts he cites are generally by authors publishing across the Atlantic.

4. Cotton Mather was by no means the only Puritan to interpret the empirical according to the scriptural, although, as I hope to demonstrate, his writing about a water-dove brings to the fore certain complications of doing so. Such interpretive acts are generally classified as typological. As Mitchell Breitwieser defines it, "Typology takes up a concrete experience of a person (including oneself), thing, or event, highlights a trait that reveals the referent's participation in a preordained and historically repetitive category, and then declares the referent's other traits (those that might make the emblematicity seem partial, unimportant, secondary, or derived) to be inconsequential for determining the referent's state of being—at best, pleasantly ornamental, at worst a blurring or obfuscating of the true. Thus typology is not antithetical to experience per se, but to those aspects of represented experience that do not confirm it" (*American Puritanism*, 24). Typology demanded an engagement with the empirical world that was limited or contingent, for much might be encountered that could not ultimately be ratified by scripture and would therefore need to be ignored. Typology fundamentally anticipates such exclusions, as Breitwieser explains: "Puritan theory is thus by design a hermeneutic violence directed against Puritanism's others" (25).

As I will argue, in a key passage Mather appears to perform such hermeneutic violence against a variety of others, but his writing actually relies upon the otherness he purports to exclude. Mather may be read as requiring engagement with Native American experience in order to arrive at his scriptural conclusion. Hence, typology cannot completely explain the oddities of Mather's text. My position responds to Michael P. Clark's recent proposal that scholars may now be in a position to think beyond typology: "After more than a half-century of relentless interrogation by scholars of American intellectual history, typology has become so closely identified with Puritan hermeneutics that it is difficult to remember, or even see on the page, the congeries of other methodological influences and exegetical practices that affected the way Puritans interpreted signs" (417). I have made an effort here to "see on the page" the traces of Native American thinking and the history of colonial violence that effect (and cannot be excluded from) what might be construed more narrowly as typological exegesis.

5. Adrienne Mayor observes that it is in the very process of giving the Native Americans authority by relating their story that he doubles back: "Mather seems to have made [the decision to deny the Native American source] in midsentence. I think he sensed that opening the door to alternative, diverse ideas about giant bones threatened his dogmatic 'science' based on a literal reading of the Bible" (*Fossil Legends*, 36–37). Mather cannot open the door of Maughkompos without shutting, to some degree, the door of Adam. But as Mayor notes, Mather seems to leave a trail of this realization

in the text, which is a manner of leaving ajar what ought to be closed—or, at least, of indicating the precarious operations that underpin his work.

6. Here and throughout my analysis, I emphasize Mather's encounters with Native American tradition as constitutive of his writing, even when he explicitly denies its pertinence. My approach is in contrast to that outlined in Jan Stievermann's introduction to a 2010 collection of Mather essays, which focuses on the "supra-regional dimensions of his thinking and writing" and the ways that he was "in dialogue with his contemporary peers across the Christian world" ("Introduction," 22, 23). I forgo the "supra-regional" aspects of Mather's work, providing a more local account, in order to bring into relief the aspects in his text that resist exegetical logic. In reading Mather's title as evocative of his colonial location, I also depart from Breitwieser's association of the "Americana" of the book's title with its structural form. See Breitwieser, "All on an American Table."

7. Mather's migrating-bird theory appears in the *Biblia Americana*, Jeremiah 8:7, as well as in the "Curiosa Americana." See Dopffel, "Between Biblical Literalism and Scientific Inquiry."

8. Abenaki vocabulary from the colonial era is difficult to reconstruct. I have used Joseph Laurent's 1884 text as well as http://www.cowasuck.org/language.cfm and http://www.bigorrin.org/archive60.htm. It may be worth noting, in the spirit of the pigeon effect, that the Abenaki words I discuss do not appear in the most recent, twentieth-century Abenaki-English dictionary (Day, *Western-Abenaki Dictionary*).

9. See Dopffel, "Between Biblical Literalism and Scientific Inquiry," 216.

10. For two related interpretive undertakings that have been playing in the background of this essay, see, on "this" in seventeenth-century religious writings, Louis Marin's *Food for Thought*, chap. 1; on "this" as applied to animal specimens, Jacques Derrida's *The Animal That Therefore I Am*, chap. 1.

WORKS CITED

Beall, Otho T. "Cotton Mather's Early 'Curiosa Americana' and the Boston Philosophical Society of 1683." *William and Mary Quarterly* 18, no. 3 (1961): 360–372.

Breitwieser, Mitchell. "All on an American Table: Cotton Mather's *Biblia Americana*." *American Literary History* 25, no. 2 (2013): 381–405.

———. *American Puritanism and the Defense of Mourning: Religion, Grief, and Ethnology in Mary White Rowlandson's Captivity Narrative*. Madison: University of Wisconsin Press, 1990.

Clark, Michael P. "The Eschatology of Signs in Cotton Mather's 'Biblia Americana' and Jonathan Edwards's Case for the Legibility of Providence." In *Cotton Mather and* Biblia Americana—*America's First Bible*

Commentary: Essays in Reappraisal, ed. Reiner Smolinski and Jan Stievermann, 413–438. Tubingen: Mohr Siebeck, 2010.

Day, Gordon M. *Western-Abenaki Dictionary*. Hull, Quebec: Canadian Museum of Civilization, 1994.

Derrida, Jacques. *The Animal That Therefore I Am*. Ed. Marie-Louise Mallet. Trans. David Wills. New York: Fordham University Press, 2008.

Dopffel, Michael. "Between Biblical Literalism and Scientific Inquiry." In *Cotton Mather and* Biblia Americana*—America's First Bible Commentary: Essays in Reappraisal*, ed. Reiner Smolinski and Jan Stievermann, 203–226. Tubingen: Mohr Siebeck, 2010.

"An Extract of Several Letters from Cotton Mather, D. D. to John Woodward, M.D. and Richard Waller, Esq.: S. R. Secr." *Philosophical Transactions (1683–1775)* 29 (1714–1716): 62–71.

Laurent, Joseph. *New Familiar Abenakis and English Dialogues*. Quebec: L. Brousseau, 1884.

Marin, Louis. *Food for Thought*. Trans. Mette Hjort. Baltimore, Md.: Johns Hopkins University Press, 1989.

Mather, Cotton. *Biblia Americana: America's First Bible Commentary, a Synoptic Commentary on the Old and New Testaments*. Vol. 1: *Genesis*. Tubingen: Mohr Siebeck, 2010.

Mayor, Adrienne. *Fossil Legends of the First Americans*. Princeton, N.J.: Princeton University Press, 2005.

Silverman, Kenneth. *The Life and Times of Cotton Mather*. New York: Harper & Row, 1984.

———. *Selected Letters of Cotton Mather*. Baton Rouge: Louisiana State University Press, 1971.

Smolinski, Reiner. "Introduction." In *Biblia Americana: America's First Bible Commentary, A Synoptic Commentary on the Old and New Testaments*, Vol. 1: *Genesis*, by Cotton Mather, 3–210. Tubingen: Mohr Siebeck, 2010.

Stievermann, Jan. "The Genealogy of Races and the Problem of Slavery in Cotton Mather's 'Biblia Americana.'" In *Cotton Mather and* Biblia Americana*—America's First Bible Commentary: Essays in Reappraisal*, ed. Reiner Smolinski and Jan Stievermann, 203–226, 515–576. Tubingen: Mohr Siebeck, 2010.

———. "Introduction." In *Cotton Mather and* Biblia Americana*—America's First Bible Commentary: Essays in Reappraisal*, ed. Reiner Smolinski and Jan Stievermann, 1–60. Tubingen: Mohr Siebeck, 2010.

Wills, David. "Meditations for the Birds." In *Demenageries: Thinking (of) Animals after Derrida*, ed. Anne Emanuelle Berger and Marta Segarra, 245–263. Amsterdam: Rodopi, 2011.

Winthrop, John. Letter to Cotton Mather. December 29, 1719. Benjamin Colman Papers. Massachusetts Historical Society, Boston.

CHAPTER 3

Smart's Professors: Birdsong and Rhetorical Agency in *Jubilate Agno*

Fraser Easton

let	whose	for
giddalti	word	the
rejoice	pairs	owl
with	with	stands
the	their	for
mocking-bird	delivery	eloquence
who	across	and
takes	the	smart
off	page	professors

Christopher Smart's long religious poem *Jubilate Agno* (1758–1763) has had a curious sort of cultural impact since it was first published in 1939.[1] Although not (yet) a canonical work, Smart's poem has exerted considerable influence on a host of modern composers and poets, from Benjamin Britten and Allen Ginsberg to Anne Sexton and Wendy Cope.[2] It has also earned significant critical and scholarly scrutiny.[3] But it is in the emerging fields of animal and posthumanist studies that *Jubilate Agno*, which has animals and their voices at its heart, has become required reading. Smart's poem pairs named individuals with a range of created species—animal, vegetable, and mineral—in an ecstatic address to God on a range of religious, autobiographical, and scientific issues, but it is best known for its much-anthologized meditation on Smart's cat Jeoffry.[4] Animal studies of the *Jubilate* have almost universally focused on this passage, interpreting Jeoffry variously as a symbolizing creature, as an emblem of Smart, of the poet, or of language, and as the key example of the creaturely voice in the poem.[5] In this essay, I want to turn attention to another key creature in Smart's poem, one which at least on the surface has more in common with the poet than an unmusical cat and his "*tribe*" (B 722): the bird and its songs.[6]

As I will demonstrate in this essay, birds and their songs in *Jubilate Agno* provide us with an essential window on the literary representation of the creaturely voice. Smart uses birdsong to portray a rhetorical agency that lies outside the domain of ideas and their associations, on the one hand, and the materiality of the signifier, on the other. Smart conceptualizes this agency in terms of deliberative speech, classically known as oratory, and he connects it to the revival of the ancient canon of delivery, dubbed elocution, in eighteenth-century England.[7] Understanding the elocutionary nature of rhetorical agency in *Jubilate Agno* and the implications of this agency for conceptualizations of the creaturely voice are the goals of my argument about Smart's birds. Along the way, we may also have a chance to rethink Jeoffry's import for animal studies.

Right from the start of *Jubilate Agno*, Smart enumerates dozens of bird species. These include species that are commonplace in literature, such as the Nightingale, the Eagle, and the Dove, as well as others that are more unusual, either because of their names, such as the Iynx, the Chloris, and the Criel; their behaviors and characteristics; or their presentation. What stands out in all these verses, however, are the ways in which Smart endows birds with rhetorical agency. Consider, for example, his satiric portrayal of "the great Owl":

> Let Ithream rejoice with the great Owl, who understandeth that which he professes.
> *For I pray God for the professors of the University of Cambridge to attend and to amend.* (B 69)

The owl, symbol of wisdom, is a stock figure for the understanding. Here, intriguingly, what "the great Owl" understands is what "he" himself "professes" or says. To profess means to declare publicly, to teach, or to affirm one's faith, among other things. We think we know that every person understands their own utterances, at least ordinarily; because of this assumption, the Let-verse serves to grant the Owl rhetorical personhood, presenting him as a self-conscious orator. In turn, the great Owl's rhetorical agency sets the stage for a reversal in the For-verse: Not *all* persons understand what they profess, even in the ordinary course of things, particularly some of those humans whose profession it is (ironically) to know things. Whether he intends to criticize their vocation as teachers or their assertion of faith, Smart shames these Cambridge "*professors*": they need "*to attend* [listen] *and to amend.*" In a spectacular put down, the verse pair presents this bird as a more competent rhetorical actor than the educators of the clergy.[8] The Owl's hoots are not emptily uttered; it is the university lectures that are unthinkingly spoken or the liturgy that is speciously recited in the university chapel. Further, in the juxtaposition within this verse pair, Smart raises the possibility that the Cambridge faculty should listen to the great Owl himself. Professors should be stirred by this bird, moved by him to give a hoot.[9]

The subtext of the verse pair is the mindful and genuine public avowal of one's faith. Cambridge in this period trained students for the clergy and, like Oxford, was commonly viewed as in decline. By calling on the great Owl to celebrate God with Ithream (the sixth son of David, the psalmist, by Eglah), Smart links the bird's rhetorical agency—his professing hoots—to poetry and music as the modes of his expression of faith.[10] Birds in poetry often exemplify song and music, and poetry itself (even if owls per se do not). Although Smart particularizes this Owl in part through his affiliation with David as psalmist, *all* the creatures in *Jubilate Agno*, human and nonhuman, are portrayed as singing to God—not just Ithream, or this great Owl, or even birds in general. Fish, too, for example, sing to God: "*For the praise of God can give to a mute fish the notes of a nightingale*" (B 24).[11] This For-verse analogizes the rhetorical agency of animals that are excluded from airborne communication—animals technically without a voice, "*mute*," such as fish—with the nightingale's music. Such privileged positioning of birds as self-aware rhetors underscores *Jubilate Agno*'s sacred purpose as a chorus of praise to God. Smart seeks to provide a work that the faithful can read aloud, together, in worship, and in so doing manifest their faith. He uses a host of devices, techniques, and themes in *Jubilate Agno*, in addition to the natural emblems of birds, to encourage readers to verbalize the poem: puns; antiphonal forms; borrowings from the Magnificat, the Benedicte, and the Psalms; foreign words; and references to versification, to name just a few.[12] Despite the poem's innovative use of long-line form and its associated prosiness,[13] these elements make it impossible for a reader (whether reading the poem aloud or not) to overlook the importance of its sonic elements and nature. Smart renders his birds as rhetorical agents whose professions are a musical analogue of persuasive speech.

Just as the great Owl serves as an emblem of the avowal of faith in *Jubilate Agno*, his hoots serve as a biographically-keyed figure for the oratorical nature of Smart's poetry and for *Jubilate Agno*. Smart was a graduate of Pembroke College, Cambridge, making the For-verse personal: He upbraids his former teachers and fellow students. Well might this one-time Cambridge scholar, confined for madness on account of a propensity (among other matters) for praying in public, wish to see the professors at his old university undergo a religious reformation (D 190). Smart clearly wants his former colleagues to "*attend*" to *his* words, as well as the owl's hoots. Verses scattered throughout *Jubilate Agno* make the case for Smart as the (unorthodox) author of a testament in which the divine word is always already a spoken one: He has "*glorified God in GREEK and LATIN, the consecrated languages spoken by the Lord on earth*" (B 6), he preaches "*the very GOSPEL of CHRIST*" (B 9), he "*sing*[s] *a psalm of* [his] *own composing*" (B 32), he is the author of his own "*MAGNIFICAT*" (B 43), and, before he was confined and began writing *Jubilate Agno*, he "*blessed God in St James's Park till* [he] *routed all the company*" (B 89). The mission of Smart's poem is to repair being through a new "*musically performed*" (B 252) liturgy; he goes so far as to ask "the Lord [to] make me a fisher of men" (B 110), that is, another apostle or Christian rhetor.[14] The verse pair under discussion applies this oratorical mission to the teachers and students of Cambridge. The great Owl, emblem of wisdom because he "understandeth that which he professes," offers utterances that are, as we have seen, greater than those of the "*professors*" of Cambridge, despite all their learning and all their books. In this sense, we can say that Smart's verse pair foreshadows something of Wordsworth's "vernal wood": "Books! 'tis a dull and endless strife, / Come, hear the woodland linnet, / How sweet his music; on my life / There's more of wisdom in it."[15]

Unlike Wordsworth, Smart is not seeking to conjure up for contemplation a vernal wood or its inhabitants. *Jubilate Agno* does not focus on an isolated subject observing nature and reacting to it within a narrative of pedagogical encounter (although some animal rhetors do come in for such consideration, like the great Owl). Rather, Smart creates a speech situation that pairs the signifying voices of men (and some women) with the signifying voices of animals, plants, and even minerals. Oratory, speaking aloud, not Wordsworthian spectatorship or auditorship, forms the Urscene of *Jubilate Agno*:

> Rejoice in God, O ye Tongues; give the glory to the Lord, and the Lamb.
> Nations, and languages, and every Creature, in which is the breath of Life.
> Let man and beast appear before him, and magnify his name together.
> Let Noah and his company approach the throne of Grace, and do
> homage to the Ark of their Salvation.
> Let Abraham present a Ram, and worship the God of his Redemption.
> (A 1–5)

Although *Jubilate Agno* was first published in 1939 from an MS that survives only in fragments, enough of it remains to show that it was composed in parallel verse columns.[16] The verses in the left-hand columns start with the word "Let," and the verses in the right-hand columns start with the word "For," creating an antiphonal structure seen, for example, in the verse pair on the great Owl. The lines just quoted, which open the poem, are missing the For-verses. But despite their loss, the remaining material shows how from the very start of his poem Smart brings together in nearly every Let-verse a named human individual to "praise" or "bless" or "rejoice" with a species of animal or plant or kind of mineral. These scenes of praise both pair human and animal speech and guide the reader towards actively sounding out the poem, rather than silently reading it. Smart conceives his poem as a canticle or liturgical psalm, even a male-authored Magnificat:

> Let Jubal rejoice with Cæcilia, the woman and the slow-worm praise the name of the Lord.
> *For I pray the Lord Jesus to translate my MAGNIFICAT into verse and represent it.* (B 43)

The Magnificat is Mary's spontaneous song of praise to God on learning of her conception of Jesus. This complex verse pair is another early declaration that the poem is meant as a rhetorical scene of speech and song, a new sonic ark of the "Testimony" (A 16) that aims to "*translate*" its reader into one of its speakers or singers.[17]

Although all the animals in *Jubilate Agno* are called on to address God, the rhetorical agency of birds holds a special place in the poem. The poem exhibits the importance of this agency in a number of ways, through thematic, structural, and aural aspects. We see it, first of all, in how Smart makes birds central to his own self-presentation, both as a person and as a poet. In verse after verse, birds reflect or define Smart's character, situation, and vocation. No other sequence of animal, plant, or mineral verses embodies autobiographical representation in the poem with the same extended intensity as the bird verses do.[18] Consider, for example, the first verse pair of the B fragment:

> Let Elizur rejoice with the Partridge, who is a prisoner of state and is proud of his keepers.
> *For I am not without authority in my jeopardy, which I derive inevitably from the glory of the name of the Lord.* (B 1)

Smart freely revises a biblical verse that associates a caged Partridge with the heart of the proud, finding in both caged bird and his own imprisonment grounds for elevation in state and in power. The caged starling of Laurence Sterne's *A Sentimental Journey* is a decade in the future; like the starling, Smart cannot get out, and he suffers (for example, from his gaolers' "*harping-irons*" [B 124]), but by associating himself with the Partridge rather than the starling, with a proudly articulate rather than a helplessly complaining creature, he turns his confinement into a moment of claimed rhetorical power via the enunciation of God's "*name*."[19]

Bird verses offer some of the most evocative and memorable autobiographical revelations in *Jubilate Agno*:

> Let Sarah rejoice with the Redwing, whose harvest is in the frost and snow.
> *For the hour of my felicity, like the womb of Sarah, shall come at the latter end.* (B 16)

The Redwing "migrates to Britain" and "feeds on berries" in the winter,[20] while Sarah bore a child in her old age; Smart links his own hoped-for future fulfillment to the late fulfillments of Sarah and the Redwing. Again:

> Let Hushim rejoice with the King's Fisher, who is of royal beauty, tho' plebeian size.
> *For in my nature I quested for beauty, but God, God hath sent me to sea for pearls.* (B 30)

Smart was a short man and hardly beautiful. By associating himself with the smallness of the King's Fisher, Smart connects his small stature to beauty as the object of an internal quest, a quest with a sublime futility about it, like a search for pearls in the ocean. Various other birds serve to map Smart's character, embodiment, and experience: the Redshank is an emblem of both his generous understanding of created things and of his personal salvation (B 11), the Iynx of his capacity for triumphant wit (B 17), the Eleos (another owl) of his singing "*a psalm of* [his] *own composing*" (B 32), the Criel of his stature (B 45), the Gull of his foolhardiness (B 51), the Fieldfare of his experience of scarcity (B 57), the Water-wag-tail of his exposure to visitors touring the asylum (B 63), the Buteo of his voice (B 80), the Oripelargus of his nobility (B 86), the Glottis of his speech (B 91), and the Glede of his fasting (B 117). Smart does link himself with other animals, such as the fish as an emblem of Christ and his cat Jeoffry as a secret sharer of God's grace, but it is with the bird verses that Smart opens the autobiographical dimension of his poem, develops it extensively, and directly ties it to the rhetorical agency of speech.

The structural dimension of the bird verses further underlines the importance of communication to *Jubilate Agno*. These verses form a coherent cluster, a sequence that concludes the A fragment and opens the B fragment.[21] From A 104 through B 122, a stretch of 132 Let-verses in total, Smart enumerates eighty-one species of birds and uses them to advance or elaborate the themes of his poem.[22] The A fragment ends with a ten-line cluster, which, as highlighted by the example of the versicles on the Nightingale and the Glowworm (A 105–106), symbolizes music and song (see Appendix). The B fragment, the longest surviving continuous segment of the poem (which concludes with the passage on Smart's cat Jeoffry), opens with seventy-two human-bird pairings across 122 verse pairs, starting with the autobiographical Partridge. Across the Let-verses of both fragments, Smart uses birds as emblems for a host of qualities. Birds symbolize music and poetry, birds express elements of Smart's autobiographical character and experience, and birds represent various physical and emotional characteristics: The Gnesion is an emblem of height (B 10), the Saurix of melancholy (B 71), and the Gier-eagle of "great penetration" (B 84). But the key quality of the bird verses is their focus on communication, with birds representing all its varieties, from the post office, letters, learning, memoirs, and translation to professing, praying, psalms, meter, music, and the voice: The Nightingale is a "musician" (A 105), the Pigeon "will carry a letter" (B 22), the Eleos is a "nightly Memorialist" (B 32), the Ossifrage is associated with "prayer" (B 54), and the Buzzard, "who is clever" (B 60), is paired with the acoustic bookishness of Bukki.[23] The three-testicled Buteo reminds Smart to "*bless God in the strength of my loins and for the voice he hath made sonorous*" (B 80), giving emblematic form to Smart's power as a poet through the association of the virile Buteo with a pun on loins/lines.

The bird verses form a structural unit not only because they are all about birds, or the person of Smart, or recurrent themes around communication, but also because these verses have an element of narrative coherence about them: They begin and end by describing how birds and their songs communicate between heaven and earth. The bird sequence opens with the Dove:

> Let Elias which is the innocency of the Lord rejoice with the Dove. (A 104)

Here Elias, "the type of John the Baptist," the forerunner of Christ, is associated with the symbol for the Holy Spirit as it descended on Jesus's baptism.[24] The first bird in *Jubilate Agno* symbolizes God's manifestation on earth. It reinforces the sense of the poem as a work with liturgical ambitions, one that "rejoice[s]" (A 104) in the Dove (the Holy Spirit) as much as the Lamb (Son).

Immediately following the verse on this most Christian of birds, Smart presents one of the most classically poetical of birds, the Nightingale:

> Let Asaph rejoice with the Nightingale—The musician of the Lord! and the watchman of the Lord! (A 105)

The nightingale is a stock figure for the poet as singer and for poetry as song, but in this and the following Let-verses from the A fragment (see Appendix), Smart specifically reworks the portrayal of a nightingale found in Andrew Marvell's poem "The Mower to the Glo-Worms." In Marvell's poem, the "Flame" of the Glo-worms "[t]o wandring Mowers shows the way"; similarly, Smart's Glowworm is "the lamp of the traveller" (A 106). But for Marvell, the Glo-worms also aid the Nightingale: "Ye living Lamps, by whose dear light / The Nightingale does sit so late, / And studying all the Summer-night, / Her matchless Songs does meditate."[25] Like Marvell's, Smart's Nightingale is a watcher as well as a singer. In Smart's poem, however, the Nightingale is a bird whose act of watching takes a characteristic of the Holy Spirit (watching over the creation) and marries it with song. Smart reimagines the connection that Marvell makes between the "living Lamps" of the Glo-worms and the Nightingale by making his Glowworm the "mead," that is, the infusing spirit or inspiration, "of the musician" (A 106), who is the Nightingale and, by extension, the poet. In this manner, Smart ties the Holy Spirit as Dove to the traditions of love poetry and Greek legend by Christianizing the Marvellian Nightingale and, indirectly, all the other avian songsters in *Jubilate Agno*. Crucially, Smart presents his Nightingale and Glowworm as conscious and effective rhetorical agents, with one lighting the scene and inspiring a religious song, the other observing and singing. In contrast, for Marvell's speaker, distracted by his lover's arrival, the Glo-worms are ultimately *in*effective and "wast" their "courteous Lights."[26]

Smart sustains this Christianizing of birdsong right through to the conclusion of the bird sequence with a verse pair on the Cherub. These lines offer a striking coda to the opening versicle on the Dove:

> Let Cherub rejoice with the Cherub who is a bird and a blessed Angel.
> *For I bless God for every feather from the wren in the sedge to the CHERUBS and their MATES.* (B 122)

Karina Williamson notes that the barn owl was sometimes called a Cherubim.[27] Whereas the Holy Spirit descends to earth in the form of a Dove (A 104), the Cherub is an earthly bird who is also an inhabitant of heaven, an Angel. The For-verse makes Smart's meaning clear: There is a great chain of bird-being that runs from the lowly wren to the angelic Cherub, a feathered ladder going *up* to heaven, just as the Holy Spirit, as a Dove, comes *down* from heaven and in Smart's account appears to gift his Christian watchfulness to the Nightingale and other birds. This narrative movement echoes a similar suturing of heaven and earth a little later in the poem: "*For the nets come down from the eyes of the Lord to fish up men to their salvation*" (B 131). Here the analogy of fishing with persuasion sustains the idea of Christ as a fisher of men. If the apostles are Christian rhetors, persuading unbelievers with the good news of the gospel, the eyes of God figuratively fulfill the same rhetorical function, physically translating persuaded men to heaven.

Along with symbolizing Smart's vocation as a poet and as a public orator in music and prayer, *Jubilate Agno*'s feathered ladder presents birds as general figures of mediation:

> Let Libni rejoice with the Redshank, who migrates not but is translated to the upper regions.
> *For I have translated in the charity, which makes things better and I shall be translated myself at the last.* (B 11)

Through the movement of their bodies in flight and in song, birds mediate between heaven and earth (Morphnus "is a bird of passage to the Heavens" [B 62]); Christian salvation means to be moved from earth to heaven, a movement that, for Smart, may also require a creature to be "*translated*" or interpreted correctly, that is, "*in the charity*," mercifully, as Christ would, and as Smart himself does, for example, "with the Crocodile, which is pleasant and pure, when he is interpreted, tho' his look is of terror and offence" (A 46). Given the highest possible thematic importance, Smart imagines birds as the agents of a rhetorical mediation between the material and spiritual realms.

Crucially, not only is the *theme* of rhetorical agency present in the bird verses of *Jubilate Agno*, but these verses *exhibit* rhetorical agency through numerous paralinguistic features that arise in the verbal performance of the poem. Paralinguistic bird verses embody or describe verbal and gestural elements such as puns, tone, articulation, volume, demeanor, and movement that can all be understood as part of the physical delivery of speech or song. Consider the deftly humorous pun in the following Let-verse:

> Let Shaul rejoice with Circos, who hath clumsy legs, but he can wheel it the better with his wings.—
> *For the banish'd of the Lord come about again, for so he hath prepared for them.* (B 39)

The name of this bird, a hawk noted for being lame in one leg, anticipates how he overcomes his disability: Lame on the ground, the Circos can "wheel it" in flight. A wheel is a circular tool used to facilitate ground travel, but this bird likes to wheel around as he flies. The power of the Circos is in the circular, and the circle in this case is once again a mediating motion, a movement that brings those out of God's favor back into it, translating them in all Smart's senses.

In other verses, Smart draws his puns from a variety of languages, especially those "*spoken by the Lord on earth*" (B 6). We see this in the case of Eleos (B 32), whose name plays on the nearly identical Greek words for owl and for mercy. Or consider the use of Latin wordplay in this verse pair on the mute swan:

> Let Shelumiel rejoice with Olor, who is of a goodly savour, and the very look of him harmonizes the mind.
> *For my existimation is good even amongst the slanderers and my memory shall arise for a sweet savour unto the Lord.* (B 3)

The Olor, with its beautiful "look," is associated with *olere*, "to savor of," through paronomasia, so that a harmony of taste is paralinguistically transferred to the bird's demeanor through a pun.[28]

Intriguingly, Smart offers a second paralinguistic dimension in the versicle on the Olor. Independently of the pun on *olere*, the bird's "look" is characterized as affecting the mind of an observer through the physical proportions of the bird. Smart's words describe rather than enact this affect, of course, but what he details is a properly rhetorical event, the bird's paralinguistic impact (through his "look") on the mind of an observer. The Olor transfers harmony to an observer through the communicative agency of its demeanor. Other birds also commune visually with observers through demeanor and movement: The Chloris communicates through "the vivacity of his powers" as well as "the beauty of his person" (A 108). The Goldfinch is "full withal" when he sings (A 109), and the Pellos is "stately" (B 112). We should think of these gestures as intentional, not simply as natural signs of internal or external states. In the case of the Olor, for example, Smart matches his own reputation in the face of slander in the For-verse to the effect of the bird's demeanor in the Let-verse, implying a deliberate reserve in the "savour" or rhetorical affect of both creatures.[29] This agency is even clearer with the Plover, "who whistles for his live, and foils the marksmen and their guns" (B 4), in two ways: First, Smart links the Plover's rapid movement with the bird's verbal articulation through the double meaning of whistle as sound and movement, and second, he casts the bird's elusiveness as an act of deliberately misleading would-be hunters. We see this rhetorical purpose in other examples of gestural communication among *Jubilate Agno*'s birds, such as the attention getting of the aptly named Water-wag-tail, who "loves to be looked at" (B 63).

Another paralinguistic capacity of birds in *Jubilate Agno* arises from the power of variation in the articulation of verbalized material. Consider the Redbird:

> Let Benjamin bless and rejoice with the Redbird, who is soft and soothing. (A 112)

The song of the Redbird is made up of the sequence of notes it produces, of course, but unlike, for example, the "Onocrotalus, whose braying is for the glory of God, because he makes the best musick in his power" (B 19), the Redbird's notes are "soft" and have a "soothing" effect on his listeners. By modulating a single element of articulation, volume, the Redbird is portrayed as generating a specific rhetorical effect on the listener, easing his or her tension. Other verses show that the rhetorical power of articulation may be implied in these verses even where it is not spelled out. For example, several versicles reference another element of articulation, the tone of the bird's song, such as the Woodlark's "sweet" sound (A 107) and the Goldfinch's "shrill" sound (A 109). Yet another element, clarity of utterance—analogous to correct enunciation—is attached to the Linnet, whose song is "distinct" (A 111). Articulation even receives thematic attention, via the physical obstacles to it: the Glottis's "tongue is wreathed in his throat" (B 91).

Jubilate Agno's birds also use another paralinguistic mode, mimicry, whether deceptive, mocking, or ironic. Consider the physical movements of the Iynx, or wryneck:

> Let Rebekah rejoice with the Iynx, who holds his head on one side to deceive the adversary.
> *For I shou'd have avail'd myself of waggery, had not malice been multitudinous.* (B 17)

Sexually insubordinate, Rebekah deceived Isaac about his children; in Greek mythology, Iynx was a nymph who charmed Zeus and, on their sexual union being found out, was punished by Hera by being transformed into a bird with a serpentine (and thus ironically phallic) neck. Smart, it seems, cannot help but laugh at the sexual undertones of his linkage of Rebekah and the Iynx, which partly accounts for the reference to "*waggery*" in the For-verse. But malice, not mirth, predominates across the verse pair, and to escape the "adversary" (devil), the bird Iynx, masculinized by Smart, deceives its opponent by moving or wagging its neck "like a snake."[30] The nymph

Iynx was sometimes described as the daughter of Echo, but rather than the repetition of words or sound, the bird Iynx imitates a gesture to escape its enemy.

The Iynx is not the only purposive mimic in Smart's bird verses; there is also a verbally parroting creature who appears early on among Smart's singers (see Appendix):

> Let Giddalti rejoice with the Mocking-bird, who takes off the notes of the Aviary and reserves his own. (A 110)

Consider first of all the reserve of Smart's Mocking-bird: He does not offer the listener his own music; rather, he holds it back. This contrasts with the Goldfinch of the preceding versicle, who is described as "shrill and loud, and full withal" (A 109). The Mocking-bird is the opposite of that little egotist, using imitation to escape from self via the repetition of the music ("notes") of others. Smart seems to embrace the Mocking-bird for his caution or even modesty in withholding his own song and instead reproducing the notes of "the Aviary." An aviary evokes an entire community of birds, maybe even a parliament of them, although here it is not speech but song ("notes") and not a legislative body but an aggregation ("Aviary") that the Mocking-bird imitates. Smart's Mocking-bird mocks in the sense that he "takes off" or reproduces the "notes" of other birds—all others, potentially. Mockery may elsewhere imply "*malice*" (B 17) or scorn, attitudes that Smart abhors in *Jubilate Agno*, but Smart celebrates what he calls "*waggery*" (B 17, 736) even in the midst of "*gravity*" (B 736), for example in the behavior of his cat Jeoffry.

Because the Mocking-bird potentially voices, through imitation, the notes of *all* birdsong, he is perhaps the birdiest of all *Jubilate Agno*'s birds.[31] But if that is the case, then are all of Smart's birds mockers, mimic-hens, secondary creatures to the men and musicians they are paired with? In other words, is there a general insight here about all birds as mimics, even if they are not literally portrayed as "tak[ing] off" the "notes" of other birds themselves? Giddalti, the man with whom the Mocking-bird is paired by Smart, was a singer in the temple in Jerusalem; as a religious singer, we can associate Giddalti along two axes: with singing birds, on the one hand, and with Smart as the author of a new liturgy, on the other. In this liturgical context, birdsong is derivative or imitative not from any deficit or secondariness, but because it explicitly repeats a set of "notes." The feathered ladder thematizes this repetition when it doubles the movement, in reverse, of the Holy Spirit descending to earth as the Dove. Elsewhere, Smart deliberately highlights the imitative nature of *Jubilate Agno* itself by referring to his work as an ark of the testimony and a new Magnificat; in short, as an imitation of the actions and words of Noah and Mary. Even when birdsong is not taking off the notes or gestures of another creature, as the Mocking-bird, Iynx, and Plover do, it does repeat *itself*, which is why we can usually identify a bird through its performance of a specific melody or sequence of notes characteristic of its species. As with birdsong, repetition is the ground on which *Jubilate Agno* as a song of praise is based: The whole point of a liturgy, even of praise, is that what is spoken or sung by the priest and the congregation is the verbal performance of a *set* formulation.

It may seem reasonable, then, to interpret the bird verses as a sequence that generalizes a Mocking-bird technology of imitation. Smart reflects extensively in *Jubilate Agno* on the powers of imitation, and these reflections connect the poem to a long tradition of thought about mimesis.[32] In the present context of the bird verses, however, neglecting the paralinguistic agency held by the imitator as rhetor in favor of a focus on the agency of signification or the sign (the notes) threatens to flatten a diverse field of acoustic phenomena onto the single plane of imitative repetition. Instead, going back to the verse pair on the great Owl and to the activity of professing as reciting a form of words (or sequence of notes) aloud, a different rhetorical technology comes into view, that of elocution, or the *verbal performance* of a set formulation.

At the same time that Smart was meditating upon a new liturgy in and around the writings that became *Jubilate Agno*, and composing the verses on birds, rhetoricians such as Thomas Sheridan sought to revive the ancient art of delivery to (among other goals) reform the Anglican church service. Elocution is the practice of effective verbal delivery of a written text or memorized script. Like Smart, Sheridan was aghast at the decline in the quality of Anglican worship. Sheridan saw the decline as a rhetorical issue, and he urged the clergy to reform their oratory through his teachings on elocution. In the context of oratory, perhaps the birdiest bird is not the Mocking-bird but the great Owl, whose act of self-conscious professing goes to the heart of *Jubilate Agno* as a liturgical performance. Considering the bird verses as a whole directs us less to a generalized technology of imitation (in the sense of semiotic iteration) than it does to a generalized technology of elocution (in the sense of gestural and verbal performance). Conceptualizing the technology of elocution has profound implications not just for how we understand the role of birdsong in *Jubilate Agno*—which is to say more generally the rhetorical agency of birds in poetry—but also for how we theorize the rhetorical agency of the creaturely voice in eighteenth-century literature.

Smart's birds can, in the aggregate, be compared to Thomas Sheridan's ideal human orator on the grounds that they make use of the same acoustic and gestural resources in their utterances. Sheridan was an eighteenth-century Irish actor, playwright, and educator who, through a series of lectures, publications, and dictionaries, popularized the need to modernize contemporary public speaking by reviving the classical canon of delivery. His *A Course of Lectures on Elocution* appeared in print in 1762 just as Smart was finishing the MS of *Jubilate Agno*, but Sheridan's ideas had already been widely disseminated: He lectured on elocution throughout Britain and Ireland before he moved to London in 1758, and he published his ideas on the subject as early as 1759.[33] In the *Lectures on Elocution*, Sheridan identifies a culture-wide decline in the standards of public communication and, in particular, a decline in the ability of literate individuals to read aloud with comfort, fluency, and comprehension (something of critical importance in the professions of law, politics, and the church). In order to reform this state of affairs, Sheridan elaborated in his lectures a technology of oral performance:

> A just delivery consists in a distinct articulation of words, pronounced in proper tones, suitably varied to the sense, and the emotions of the mind; with due observation of accent; of emphasis, in its several graduations; of rests or pauses of the voice, in proper places and well measured degrees of time; and the whole accompanied with expressive looks, and significant gesture.[34]

For Sheridan, acoustic elements, combined with countenance and gesture, make up the "articulate sounds and tones" (7) of speech through which speakers can best communicate an author's (or their own) words to someone else. As we have seen, as rhetorical agents, Smart's birds make use of exactly the same acoustic and gestural resources that Sheridan enumerates, including distinct articulation, volume, tone, countenance, and gesture. Smart's birds are little elocutionists, and the elocutionary technology of reading aloud is key to Smart's portrayal of the rhetorical agency of their songs.[35]

Like *Jubilate Agno*, Sheridan's account of elocution is also meant to be socially significant. It not only seeks to help excluded individuals overcome the cultural disabilities of regional or social accents, but more ambitiously it aims to reform the speaking professions such as law and politics. Sheridan's intervention shares a primary target with Smart's: the decline in the quality of religious worship in the established church. As Sheridan puts it in the *Lectures on Elocution*, "the general manner of reading the Church Service"

> is often so ill performed, that not only the beauty, and spirit of the service is lost, but the very meaning is obscured, concealed, or wholly perverted. I have heard many clergymen, who did not read one single sentence as it should be, from the beginning to the end. . . . And on this account it is, that there is no composition in the English tongue, which is at all attended to, so little understood, in general, as the Church Service. (59)

Sheridan charges that if the English clergy badly pronounce the liturgy and even their own sermons, the minister's "very meaning is obscured, concealed, or wholly perverted." Smart, too, is focused on the oral delivery of the liturgy, and it is not just through the theme of praise but through its performance as congregational oratory that he seeks to renew the agency of its creaturely practitioners, pairing birdsong, for example, with human voices. It is to instruct the teachers of the clergy as lecturers that Smart waggishly directs his former professors "*to attend and to amend*" (B 69) under the aegis of an Owl.

Sheridan's most disruptive claim in his lectures is that the "articulate sounds and tones" (7) of speech amount to a language without words. Sheridan agrees with John Locke that the right use of words is crucial to the communication of ideas. But Sheridan argues that Locke's focus on the representation of ideas through words ignores two fundamental features of effective communication. First of all, through pauses, emphasis, and tone, the act of reading aloud mediates the semantic dimension of a written script and co-creates the ideas that are verbally communicated. To demonstrate this, Sheridan gives the example of the variation of emphasis in the following sentence: "Shall you ride to town to-morrow?" (58). Emphasize "shall," and the sentence is made into a question about whether the addressee is determined to ride or not; emphasize "you," and it becomes a question about whether the addressee or another person will ride; emphasize "ride," and it becomes about the mode of conveyance; emphasize "town," and it becomes about destination; and emphasize "to-morrow," and it becomes about timing. Just as a change in accent may change the meaning of certain words, so a change in emphasis may change the meaning of a sentence and, in the absence of a set of diacritical marks going beyond ordinary punctuation, the mediation of words by emphasis will not be wholly determined by words or their grammar (whether in the form of written markings or as a memorized script).

The second feature of rhetorical agency that Sheridan feels Locke overlooks springs from the sentimental observation that humans do more than communicate ideas: they communicate inner states of feeling, or sentiments. The experiences of passion, feeling, and other inner states such as mood can only be *named* with words, not *conveyed* with them. Anger, for example, is the name of a mental state, not the communication of that mental state. To communicate the passion of anger, other "marks" such as tone are needed because, as Sheridan puts it, "the passions and the fancy have a language of their own, utterly independent of words, by which only their exertions can be manifested and communicated" (x). The language of passion operates, for Sheridan, through a nonword language of tones, looks, and gestures such as cries and tears to convey directly the experience of the passions from one individual to another. The promise of elocution, then, is not just the more effective communication of words and their associated ideas through a pedagogy of vibrant reading aloud but also the thoughtful communication of mental states that words cannot transmit.[36] The way words are read aloud, the tone in which they are read, and even the interjections, cries, glances, and gestures that go beyond the words being read aloud, all serve to communicate the passions of the speaker and so bind the listener to the speaker through shared feelings.

For Sheridan, then, tones are hugely consequential. Tones are "the language of the animal passions of man" (101), the basis of human and animal society, and the source of a shared communion between humans and animals. "As words are marks of ideas, so are tones of energies and affections of the mind" (73) and of "our feelings" (74); tones are "fixed, self-evident, and universally intelligible" and are part of human being in its "animal state" (101); indeed, "the tones expressive of sorrow, lamentation, mirth, joy, hatred, anger, love, pity, &c. are the same in all nations" (101). All humans, according to Sheridan, share the language of tone, and the resources of feeling and tone, not of ideas and words, found society:

> But tho' it be not necessary to society, that all men should know much; it is necessary that they should feel much, and have a mutual sympathy, in whatsoever affects their fellow creatures. (101)

Nor is it just human creatures that commune through tones:

> [A]ll animals that are not mute, or wholly incapable of uttering any sound . . . also express their passions by certain tones. . . . But it is to

> be observed, that each species of animals, seem to have a language of their own, not at all understood, or felt by the rest. The lowing of the cow affects not the lamb; nor does the calf regard the bleating of the sheep. (102)

Although animals cannot communicate across species, except through "antipathy," as when "the cry of dogs, warns the hare of his danger" (103), no "animal [can] utter any sound which we [humans] cannot explain, or tell from what emotion, or passion it proceeds" (104). Smart, in *Jubilate Agno*, seems to echo this encapsulation of animal communication when he elaborates a scheme of animal "simples" or sonic atoms behind different human languages: cat for Greek, mouse for Latin, and dog and bull for English.[37]

Jubilate Agno also echoes another idea that Sheridan discusses here, the privileged position of humans with respect to the communication of emotion. Sheridan argues that man

> is an abstract of all animal nature; and that in his tribe are to be found, all the emotions and passions, that belong to all the several tribes. (105)

Just as the language of tones is not restricted to specific language groups, it is not restricted, for humans, to human tones. Animals and humans occupy, for the latter, a single community of emotional sympathy. Intriguingly, Sheridan's use of the word "tribe" for animal communities echoes Smart's usage in *Jubilate Agno*. Smart identifies animals with various tribes, such as Jeoffry, who "*is of the tribe of Tiger*" (B 722). When therefore Smart's poem brings individual men (and some women) together with animal species (tribes) to pair their voices, it seems as if he is riffing on Sheridan's idea of the emotional universality of the vocal capacity of tone in human oratory. Like other eighteenth-century sentimental thinkers and authors, Smart levels human and animal in the field of sympathetic communication, but he may be the only one to have imagined sympathy in terms of a collective oration by all the tribes, human, animal, plant, and mineral.

To strengthen his claim that "there is no passion or emotion whatsoever, in the whole animal world, which is not to be found in man," Sheridan asserts:

> The roaring of the lion, is not more terrible than the voice of his [that is, man's or humankind's] anger; nor the cooings of the pigeon, more soft, than the murmurs of his love. The crowing of the morning cock, is not so clear and sprightly as the notes of his joy; nor the melancholy mournings of the turtle [dove], so plaintive as those of his woe. (103)

In line with other sentimental thinkers, Sheridan equates specific animal and human emotions, just as Smart does with Jeoffry. We are reminded, too, of how the bird verses delineate Smart's autobiographical feelings and concerns and of how they highlight his sense of personal rhetorical agency. Besides the examples of bird verses I have already reviewed, which use tone to communicate emotion, there are many verses in which birds are associated with specific passions and feelings: The Gull is happy (B 51), and the Saurix is melancholy (B 71); the Eleos is merciful (B 32), and the Vultur is fierce (B 72); the Blackbird is worshipful (A 113), and the Water-wag-tail is self-satisfied (B 63). However, it is not clear what weight we should place on the emblematic understanding of specific birds; in Sheridan's case, there is a strict emblem-tone identification at work that Smart studiously avoids.

With Sheridan, the equation of emblem and tone may cast the authenticity of the creaturely voice, at least for birds, in doubt: Do we really think that every hoot of a turtledove, whatever its modulation, is an expression of inner sadness? If we did, it would hard to "*consider*" (B 695) animal expression as closely as Smart does. Sheridan magnifies the issue of authenticity when he notes "the power of musical imitations" (101) of tones (and he mentions this as evidence for the independence of tones from words). Animals can certainly imitate tones, as Smart's Mocking-bird exemplifies, but given that musical instruments have no inner life or passions, what does it mean that they can be used to imitate tones and move listeners accordingly, with "power," to boot? What affect could such a tone be the sign of? As if to address this issue, Sheridan tells a story, recounting the seventeenth-century actor Thomas Betterton's riposte to a question from the bishop of London. The bishop had asked why theatrical performances move audiences more than "discourses from the pulpit, upon subjects of the utmost importance to them." Betterton is said to have replied: "'My Lord, it is because we are in earnest'" (127). Sheridan encourages us to take birdsong in earnest, as feeling and music, even as its tones are not natural signs but conventionalized acoustic marks that exceed animal corporality and, as such, are imitable. And the reason for this encouragement is clear enough: When Sheridan compares "musical imitations" to tones, he is not comparing spoken tones to the notes of a musical score (which would be like equating tones to written words); rather he is comparing tones to the performance of the score, which mediates specific notes. If the performers are in earnest, then their performance will be too, but it will not be the direct expression of "natural affect signs" inasmuch as tone is the performance of a note.[38]

What emerges from this consideration of Sheridan and Smart is a new understanding of the creaturely voice, one that is elocutionary and thus verbally performed; one that shares a common articulation in humankind, animals, and other creatures, binding them socially; and one for which tones and the musical aspect of speech is a language outside of words.[39] In this context, the voice is not programmed by signs or the inner logic of signification; it is driven by neither signifier nor signified, written or spoken grammar. This is true even when the signs that a creature voices are affect signs that derive from the physical expression of inner passions (and thus mediate those passions). Consider this line from Smart's poem:

> *For the VOICE is from the body and the spirit—and is a [sic] body and a spirit.* (B 239)

In a recent overview of eighteenth-century poetry, one prominent critic equates this For-verse to Pope's idea of the sound (body) echoing the sense (spirit) in poetry.[40] Sheridan, for his part, is clear that it is tone, rather than any script of speech, that gives "spirit" (68) to the voice (that is, to the performance of speech). Sheridan complains about "the fetters of measure" (77), the meter, which is a feature of the script of a poem, as a constraint on successful elocution.[41] Indeed, it is not too much to say that the idea of the sound echoing the sense is the very opposite of the claims of elocution, inasmuch as sounding out determines semantics for Sheridan. And in this regard what is true of Sheridan's lectures is also true of Smart's poem, whether in terms of the poem's wordplay or its complicated antiphonal structure, both of which work to let reading aloud drive sense.

Perhaps a better way to understand Smart's verse on "*VOICE*" (B 239) is in terms of wordplay on the aspirate sound and the role of the breath in speech, where the "*body*" of the vocal apparatus is moved by the "*spirit*" of breath in the performance of speech.[42] As Smart writes in the *Jubilate*:

> *For Action and Speaking are one according to God and the Ancients.* (B 562)

In this line Smart references the classical notion that delivery is language in action, that is, performed speech.[43] In this context we must also differentiate the elocutionary understanding of voice from a stylistic understanding of it. Another prominent critic describes the "poetic voice" in *Jubilate Agno* as owing its "power" to "strange diction, the praise of God, [and] a suffering condition."[44] This is a compelling analysis, under the rubric of voice, of the *ethos* of the implied author of *Jubilate Agno* as manifested through his decisions about diction and other formal and thematic elements. But the idea of "*VOICE*" that I have conceptualized in this essay is a rhetorical agency that arises *only* in speech—in the verbal performance of a text. It is a *physical* style that is not wholly captured by the semantic, semiotic, or grammatical elements of meter, diction, scheme, sign, figure, or theme.[45]

Verbalization is always implicit in the "mission" of poetry as such. Reading aloud and the expression of feeling through tones are thematized in *Jubilate Agno* in the call to universal prayer of man and creature. But as with elocutionists like Sheridan, the mediation of the voice in *Jubilate Agno* is not limited to the agency of the signifier or the "grain of the voice" (the organic support of the sign in the body); it includes the power of meaning making in its own right, via the verbal performance (delivery) of the script of (written) language. Reading *Jubilate Agno* aloud means remediating the text of Smart's poem and all the animal and human voices (cries, words, songs, etc.) silently waiting on the page to be spoken—and, as with the great Owl, understood. Smart's birds use tone to express inner emotions just like other animals—and humans. The tones of their voices (along with physical gestures and looks), then, mediate notes and words paralinguistically to communicate feeling and to bind them with the rest of the creaturely world in sympathetic society.[46]

The possibility of a sympathetic social bond between different animal species through verbal communication is one of Smart's key divergences from Sheridan. For Sheridan, different species can verbally communicate only "antipathy" (103) to one another, such as when a dog's bark frightens off a hare. Smart knows that society is not always sympathetic, a fact he acknowledges throughout *Jubilate Agno* and exemplifies in the account of his own confinement for disorderly public prayer:

> *For I blessed God in St James's Park till I routed all the company.*
> *For the officers of the peace are at variance with me, and the watchman smites me with his staff.* (B 89–90)

Smart makes the threat of violence a part of the bird verses, too: For the first eighty-two lines of the B fragment, Smart alternates references to birds with references to the creatures that birds prey upon or which prey upon them.[47] Within this juxtaposition of predator and prey, Smart draws on verbal wordplay to invoke cross-species sympathy, bringing birds, the prey of birds (insects, worms), and creatures that prey upon birds (snakes) together as rhetorical agents to celebrate God: Preying becomes praying.[48]

Prayer is an important speech genre in the elocutionary rhetoric of the bird verses; there are many others. Birds "glorify," "preach," "bless," "sing," "trumpet," "call," "scorn," "implore," "give," "attribute," "worship," "council," "throw," and, of course, "profess." Most of all, however, birds "rejoice." Rejoicing is a key speech genre in Smart's poem: *Jubilate Agno*, after all, is

Latin for "rejoice in the lamb," and the poem opens with the invocation to "Rejoice in God, O ye Tongues" (A 1). Except for a single versicle early in the A fragment (A 8), the Let-verses have men and creatures "blessing" and "praising" God until the Dove comes to "rejoice" (A 104). From A 104 to B 122, all but three of the verses "rejoice" in God.[49] The action of rejoicing, and the introduction of joy into the worship of God, are identified with the birds of the poem and their songs. A similar joy infuses Jeoffry and his worship of God. He "*kiss*[es]" (B 714) other cats, lets "*one mouse in seven*" go free (B 716), and "*is an instrument . . . to learn benevolence upon*" (B 727)—at least when he is not indulging in "*waggery*" (B 736). Like the Nightingale, he keeps "*watch in the night*" (B 718). He also counteracts "*death*" (B 720) with his "*electrical*" (B 719) powers, which, like the Glow-worm's, cast "*light about him*" (B 761). And Jeoffry is a creature of the dawn chorus, singing "*morning orisons*" in which "*he loves the sun*" (B 721), as if his prayer was an aubade to a nature that "priketh hem."[50]

Smart's reworking of Marvell's "The Mower to the Glo-worms," as it plays out across the bird verses, makes the joyful, even erotic, nature of birdsong stand as an example of the elocutionary technology of all creaturely voices. When Smart invites us to "*consider*" his cat, he draws our attention not only to Jeoffry's joy, or to its cross-species power, but to his rhetorical agency: "*For he is tenacious of his point*" (B 735). Like the birds, Jeoffry is a complete elocutionist. He uses gesture, "*wreathing his body*" (B 698) to worship and "*look*[ing] *up for . . . instructions*" (B 711) from his master; he "*spit[s]*" (B 725) when provoked and "*purrs*" (B 726) when grateful, and so uses tones; he "*can learn*" (B 744) and he can teach (B 727, 755); and he reciprocates the actions of others as "*he can fetch and carry*" (B 746), and "*he can catch the cork and toss it again*" (B 750). Jeoffry truly is one of Smart's professors, even appearing in the bird verses:

> Let Shephatiah rejoice with the little Owl, which is the winged Cat.
> *For I am possessed of a cat, surpassing in beauty, from whom I take occasion to bless Almighty God.* (B 68)

The little Owl is not literally a cat with wings, but Jeoffry's close association with it further underscores the elocutionary nature of his rhetorical agency. He even participates in the rhetorical mediation of heaven and earth through a feline version of the feathered ladder found in the bird verses: "*For the Cherub Cat is a term of the Angel Tiger*" (B 723).[51]

Jubilate Agno famously renders Jeoffry as a sort of universal creature. Smart describes Jeoffry as a particularly accomplished "*quadrupede*" (B 765) who can also "*swim*" (B 767) like a fish and "*creep*" (B 768) like a snake or an insect. It is fitting, then, to find him associated with birds and birdsong from early on in the poem. Even if "*he cannot fly*," Jeoffry is "*an excellent clamberer*" (B 764), and even if his "*tongue*" is lacking in "*musick*" (B 743), he can use gesture to "*tread to all the measures upon the musick*" (B 766). Jeoffry's rhetorical abilities ensure his full participation in the agency of speech as we have examined it in this essay. By the first of the fish verses (B 123), the universal extent of elocutionary technology is fully established. In Smart's rhetorical cosmos, all creatures, not only birds, verbally "rejoice": Even "*a mute fish*" may take up "*the notes of a nightingale*" (B 24), and rhetorical agency, equally manifest in cat and human, fish and fowl, extends to every creature.

Appendix: "Of Birdsong"

JUBILATE AGNO, A 104–113

Let Elias which is the innocency of the Lord rejoice with the Dove.
Let Asaph rejoice with the Nightingale—The musician of the Lord! and the watchman of the Lord!
Let Shema rejoice with the Glowworm, who is the lamp of the traveller and mead of the musician.
Let Jeduthun rejoice with the Woodlark, who is sweet and various.
Let Chenaniah rejoice with Chloris, in the vivacity of his powers and the beauty of his person.
Let Gideoni rejoice with the Goldfinch, who is shrill and loud, and full withal.
Let Giddalti rejoice with the Mocking-bird, who takes off the notes of the Aviary and reserves his own.
Let Jogli rejoice with the Linnet, who is distinct and of mild delight.
Let Benjamin bless and rejoice with the Redbird, who is soft and soothing.
Let Dan rejoice with the Blackbird, who praises God with all his heart, and biddeth to be of good cheer.

NOTES

1. Smart's poem first appeared under a different title and in a different format: Smart, *Rejoice in the Lamb*, ed. Stead. I would like to thank Tom Keymer, Camie Kim, Alice Kuzniar, and John Savarese for their comments on an earlier version of this essay.
2. Williamson, "Surfing the Intertext."

3. See, for example, Hartman, "Christopher Smart's 'Magnificat'"; Guest, *A Form of Sound Words*; Hawes, *Mania and Literary Style*; Gigante, "Smart's Powers"; and Menely, *The Animal Claim*, 133–147.

4. Smart, *Jubilate Agno*, ed. Williamson, B 695–768. Subsequent references to this work will appear parenthetically in the text.

5. See, for example, Shell, "The Family Pet," 121; Hartman, "Christopher Smart's 'Magnificat'"; Karremann, "Human/Animal Relations in Romantic Poetry," 102–107; Ozarska, "The Presentation of the Animal Sphere in Christopher Smart's *Jubilate Agno*," 70–72; Menely, *The Animal Claim*, 137–138; and Ozarska, "What If Christopher Smart's Cat Responded?"

6. Smart describes Jeoffry's "*tongue*" as "*want*[ing] *in musick*" (B 743).

7. The five canons of classical rhetoric were invention, arrangement, style, memory, and delivery. Delivery was revived as elocution in eighteenth-century England; this should not be confused with the Latin name for the canon of style, *elocutio*, which referred to figures of speech.

8. Keymer describes the verse pair on the great Owl as "one of *Jubilate Agno*'s more blatantly satirical combinations of versicle and response" ("Presenting Jeopardy," 97).

9. On the addressivity of the creaturely voice in this period, see Menely, *The Animal Claim*.

10. For an alternative reading of the verse pair that sets the owl as a "symbol of wisdom" against the "scientific notions" of Cambridge professors, see Ozarska, "The Presentation of the Animal Sphere," 68. While it is my argument here that it is the *avowal* of faith that the owl verse pair emphasizes and seeks to "*amend*," Smart does elsewhere specify the need for academic *moral* reformation: "I refer the people of both Universitys to the Bible for their morality" (D 190).

11. As do plants: "Let Hassenaah rejoice with the White Beet. God be gracious to Hasse and all musicians" (C 90). (The pun on Beet/beat conjures up the musicians of the second half of the versicle.) Even minerals sound out to God: "Let Close, house of Close rejoice with Chalcophonos a gem sounding like brass. O all ye gems of the mine bless ye the Lord, praise him and magnify him for ever" (D 73).

12. For influential readings of the play of signifier and signified in *Jubilate Agno*, and the relationship between the divine word and the material signifier, see Hartman, "Christopher Smart's 'Magnificat'"; and Liu, "Christopher Smart's 'Uncommunicated Letters.'" Menely identifies this interruption of "customary significance" with the creaturely voice in *Jubilate Agno* (*The Animal Claim*, 142). In contrast, I am interested in the ways in which oratory is a source for this play of signification in Smart's poem and

therefore an occasion for the agency of specific creatures in verbal performance.

13. Smart's use of long-line form is indebted to his friend Robert Lowth's *Lectures on the Sacred Poetry of the Hebrews* (trans. 1787), first published in Latin in 1753.

14. See Guest, *A Form of Sound Words*, for an account of the religious aims of *Jubilate Agno.*

15. Wordsworth, "The Tables Turned," ll. 21, 9–12.

16. As first brought to light in Bond's edition of *Jubilate Agno.*

17. On the poem as sung canticle or psalm, see Williamson, "Introduction," xxiv–v; and Walker, "'Jubilate Agno' as Psalm."

18. The fish verses (B 123–295) that follow those on birds open with thirty autobiographical lines across the first thirty-four verses of the sequence; Smart then moves on to scientific and other matters for 136 of the remaining 139 verses. Given the importance of common-place book, diary, and personal apology to the form and substance of *Jubilate Agno* as a whole, the deep relationship that Smart establishes between autobiography and birds is an indication of the centrality of the bird verses to the poem.

19. David compared "his prosecutor" to a partridge hunter (Guest, *A Form of Sound Words*, 131); this association defines Smart's authority in terms of the persecution of a public poet, that is, in terms of the vicissitudes of rhetorical agency.

20. Williamson, ed., *Jubilate Agno*, B 16n.

21. The bird verses thus bridge the poem's first entirely lost section, one that Williamson conjectures may have contained anywhere from 113 to 295 verse pairs ("Introduction," *Jubilate Agno*, vol. 1 of *The Poetical Works*, xxiii). It is likely, given Smart's well-documented tendency in *Jubilate Agno* to cluster themes, creatures, and individuals, that there were dozens, if not hundreds, of now-lost bird verses in these missing lines. For recent work on clusters in *Jubilate Agno*, see Gigante, "Smart's Powers."

22. There are around thirty other bird verses scattered outside the main cluster; see, for example, B 682–686, C 41, 82, 141, and D 187–189, 191–194, 196–199, 201, 208. These additional bird verses bring the total number of Let-verses dedicated to birds in *Jubilate Agno* to over one hundred.

23. "'Bukki' . . . is pronounced 'book-y'" (Guest, *A Form of Sound Words*, 150).

24. Williamson, ed., *Jubilate Agno*, A 104n.

25. Marvell, "The Mower to the Glo-Worms," ll. 9, 10, 1–4.

26. Marvell, "The Mower to the Glo-Worms," l. 13.

27. Williamson, ed., *Jubilate Agno*, B 122n.

28. Williamson, ed., *Jubilate Agno*, B 3n.

29. Clement Hawes discusses Smart's sense of his public reputation in the For-verse of B 3, in *Mania and Literary Style*, 174, but he does not address the bird's rhetorical agency.

30. Williamson, ed., *Jubilate Agno*, B 17n.

31. The Mocking-bird is also a fit image for Smart as an aggregator of the voices of other creatures.

32. As I argue in "Christopher Smart's Cross-Dressing," 214–215, *passim*.

33. Spoel, "Rereading the Elocutionists," 58; Howell, *Eighteenth-Century British Logic and Rhetoric*, 233–234.

34. Sheridan, *A Course of Lectures on Elocution*, 10. Subsequent references to this work will appear parenthetically in the text.

35. For an overview of elocution in *Jubilate Agno*, see Easton, "Christopher Smart's Elocution."

36. Sheridan here is suturing the period's sentimental semiotics of the affect sign to the classical rhetorical canon of delivery. The result of Sheridan's suture is that the cry of nature does not exist in isolation from a system of conventional or linguistic or musical signs but (whatever it might signify in isolation) occurs on the occasion of conventional signification, which it mediates in performance. Significantly, by pairing human and nonhuman voices, Smart guarantees, as Sheridan does not, that the affect sign will always already inhabit the linguistic sign or musical note. For a recent account of the sentimental view of the affect sign in the eighteenth century, see Menely, *The Animal Claim*, 42–79.

37. Hartman, "Christopher Smart's 'Magnificat,'" 91.

38. It is fair to say that Sheridan at times makes a straightforward sentimental case for tones as affect signs of creaturely passions, but, by engaging with sentimental ideas about affect through the classical canon of delivery, he levers those signs out of a biosemiotic (and its conventionalization) and into the space of performance. There, in performance, tone serves within oratory as an energy mediating a text through its delivery.

39. For an alternative approach to the creaturely voice in *Jubilate Agno*, see Menely, *The Animal Claim*, 124–147. Menely is interested in the metaphysics of voice, that is, of addressivity as such, and he reads Smart's poem in part as an expression of creaturely address through "the voice itself—not yet differentiated into the differential units of the linguistic sign" (143) and partly through "the creaturely imaginary" (141) as the source of the materiality of the sign. The prelinguistic space of "an address and an enunciation" without a "discernable statement" is explored by Mladen Dolar, among other thinkers influenced by Derrida and Lacan, in *A Voice and Nothing More*, 28. Smart certainly is interested in the metaphysics of the voice, as I argue in "Christopher Smart's Cross-Dressing" and "'Mary's Key' and the Poet's

Conception." But he is also interested in the *praxis* of the voice, and it is the implications of that interest for creaturely expression as rhetorical agency that is the subject of the present essay.

40. Sitter, *The Cambridge Introduction to Eighteenth-Century Poetry*, 7. The relevant line from Alexander Pope's *An Essay on Criticism* is "The sound must seem an Echo to the sense," l. 365.

41. Again, contra Pope: "But most by Numbers judge a Poet's song" (*An Essay on Criticism*, l. 337).

42. As I have argued elsewhere: Easton, "Christopher Smart's Elocution," 74.

43. For a different approach to the "vocative nature" of *Jubilate Agno*, one focused on the addressivity of the figure of the prophet-poet in Smart's poem rather than on delivery in general, see Katz, "'Action and Speaking Are One,'" 51. Gigante ("Smart's Powers," 75) and Menely (*The Animal Claim*, 146) are examples of recent critics who read this line in terms of the creative word or Christian Logos. As Gigante puts it: "In the philosophical tradition of logos guiding this poem, to pronounce is to produce." In contrast, I see a secularizing, classical idea of delivery at work in this verse.

44. Keymer, "Presenting Jeopardy," 106.

45. For other work on speech and voice in *Jubilate Agno*, see Friedman, "The Cosmology of Praise"; Mandel, "Theories of Voice in Eighteenth-Century Poetry"; and Kumbier, "Sound and Signification in Christopher Smart's *Jubilate Agno*." On puns, see Hartman, "Christopher Smart's 'Magnificat,'" 82ff.; and Guest, *A Form of Sound Words*, 175–181.

46. Menely follows sentimental thinkers of the eighteenth century in seeing natural signs of passion, affect signs, such as cries of pain—the creaturely voice—as at the origin of symbolic or conventional language and as still operative within it. The role of affect signs as primary signifiers, and their addressivity, places the creaturely voice inside language as such, in the continuing significance of nonsymbolic signifiers. Menely takes the materiality of language as the trace of the creaturely voice or signifier and so reverses Derrida's understanding of voice as pure presence. Here I am arguing instead that Smart mobilizes, within a discourse of sonicness, another creaturely property, the rhetorical agency of a particular, situated address. Particular cries are situated, oriented, and articulated in ways that are creaturely in action. This is what is missed in other interpretations of Smart's assertion that "*Action and Speaking are one*" (B 562). Absent rhetoricity, affect signs may as well be soliloquies. Rhetoricity is another name for the action of the creature on language. The creaturely voice I am interested in is this rhetorical one, the result of a mediation of a script through

performance in light of a goal or a purpose. Smart does not only unleash the passion of the creaturely signifier or advocate for the animal claim; he also specifies and exhibits the rhetorical agency of the creature.

47. This alteration is a ratio of two bird verses to every one nonbird verse and is sustained for most of the first eighty-two lines of the B fragment. From B 83 to B 122, the ratio of bird to nonbird verses varies.

48. The pun is Smart's: "Let Bedan rejoice with Ossifrage—the bird of prey and the man of prayer" (B 54).

49. From A 104 to B122 all the birds rejoice with only three exceptions: Eleos the memorialist (B 32), Saurix the melancholy (B 71), and Drepanis the passenger (B77) all "bless." That's 129 rejoicings, a large number.

50. Chaucer, "General Prologue," l. 11. For recent examples of emblematic understandings of Jeoffry's expressivity, see Ennis, "Christopher Smart's Cat Revisited," 14; Ozarska, "The Presentation of the Animal Sphere in Christopher Smart's *Jubilate Agno*," 70; and Karremann, "Human/Animal Relations in Romantic Poetry," 106. Karremann identifies a "creaturely poetics" in the "semiotic competence" of animals (97), but in examining Jeoffry's communicative acts she merely analogizes his "feline movements" with "words" (105). None of these studies takes a rhetorical approach to the creaturely voice as it is manifested in *Jubilate Agno*.

51. Compare the last versicle in the bird sequence: "Let Cherub rejoice with the Cherub who is a bird and a blessed Angel" (B 122).

WORKS CITED

Chaucer, Geoffrey. "General Prologue." In *The Complete Poetry and Prose of Geoffrey Chaucer*, ed. John H. Fisher, 9–24. New York: Holt, Rinehart, and Winston: 1977.

Dolar, Mladen. *A Voice and Nothing More*. Cambridge, Mass.: MIT Press, 2006.

Easton, Fraser. "Christopher Smart's Cross-Dressing: Mimicry, Depropriation, and *Jubilate Agno*." *Genre* 31, nos. 3–4 (1998): 193–243.

———. "Christopher Smart's Elocution." In *Reading Christopher Smart in the Twenty-First Century*, ed. Min Wild and Noel Chevalier, 63–84. Lewisburg, Penn.: Bucknell University Press, 2013.

———. "'Mary's Key' and the Poet's Conception: The Orphic versus the Mimetic Artist in *Jubilate Agno*." In *Christopher Smart and the Enlightenment*, ed. Clement Hawes, 153–175. New York: St. Martin's Press, 1999.

Ennis, Daniel J. "Christopher Smart's Cat Revisited: *Jubilate Agno* and the *Ars Poetica Tradition*." *South Atlantic Review* 65 (2000): 1–23.

Friedman, John Block. "The Cosmology of Praise: Smart's *Jubilate Agno*." *PMLA* 82 (1967): 250–256.

Gigante, Denise. "Smart's Powers: *Jubilate Agno*." In *Life: Organic Form and Romanticism*, 49–105. New Haven, Conn.: Yale University Press, 2009.

Guest, Harriet. *A Form of Sound Words: The Religious Poetry of Christopher Smart*. Oxford: Clarendon, 1989.

Hartman, Geoffrey H. "Christopher Smart's 'Magnificat': Towards a Theory of Representation." In *The Fate of Reading and Other Essays*, 74–98. Chicago: University of Chicago Press, 1975.

Hawes, Clement. *Mania and Literary Style: The Rhetoric of Enthusiasm from the Ranters to Christopher Smart*. Cambridge: Cambridge University Press, 1996.

Howell, Wilbur Samuel. *Eighteenth-Century British Logic and Rhetoric*. Princeton, N.J.: Princeton University Press, 1971.

Karremann, Isabel. "Human/Animal Relations in Romantic Poetry: The Creaturely Poetics of Christopher Smart and John Clare." *European Journal of English Studies* 19 (2015): 94–110.

Katz, Edward Joseph. "'Action and Speaking Are One': A Logological Reading of Smart's Prophetic Rhetoric." In *Christopher Smart and the Enlightenment*, ed. Clement Hawes, 47–66. New York: St. Martin's Press, 1999.

Keymer, Thomas. "Presenting Jeopardy: Language, Authority, and the Voice of Smart in *Jubilate Agno*." In *Presenting Poetry: Composition, Publication, Reception*, ed. Howard Erskine-Hill and Richard A. McCabe, 97–116. Cambridge: Cambridge University Press, 1995.

Kumbier, William A. "Sound and Signification in Christopher Smart's *Jubilate Agno*." *Texas Studies in Language and Literature* 24 (1982): 293–312.

Liu, Alan. "Christopher Smart's 'Uncommunicated Letters': Translation and the Ethics of Literary History." *Boundary 2* 14, nos. 1–2 (Fall 1985–Winter 1986): 115–146.

Mandel, Eli. "Theories of Voice in Eighteenth-Century Poetry: Thomas Gray and Christopher Smart." In *Fearful Joy*, ed. James Downey and Ben Jones, 103–118. Montreal: McGill-Queen's University Press, 1974.

Marvell, Andrew. "The Mower to the Glo-Worms." In *The Poems and Letters of Andrew Marvell*, 3rd ed., ed. H. M. Margoliouth and Pierre Legouis, 1:47–48. Oxford: Clarendon, 1971.

Menely, Tobias. *The Animal Claim: Sensibility and the Creaturely Voice*. Chicago: University of Chicago Press, 2015.

Ozarska, Magdalena. "The Presentation of the Animal Sphere in Christopher Smart's *Jubilate Agno*." *Lublin Studies in Modern Languages and Literature* 26 (2002): 63–73.

———. "What If Christopher Smart's Cat Responded?—A Human-Animal Studies Perspective on *Jubilate Agno*'s 'Cat Jeoffry.'" *Humanimalia* 8 (2016): 35–52.

Pope, Alexander. *An Essay on Criticism*. In *Alexander Pope: Selected Poetry and Prose*, ed. William K. Wimsatt, Jr., 63–84. New York: Holt, Rinehart, and Winston, 1965.

Shell, Marc. "The Family Pet." *Representations* 15 (1986): 121–153.

Sheridan, Thomas. *A Course of Lectures on Elocution*. London, 1762.

Sitter, John. *The Cambridge Introduction to Eighteenth-Century Poetry*. Cambridge: Cambridge University Press, 2011.

Smart, Christopher. *Jubilate Agno*, ed. W. H. Bond. London: Rupert Hart-Davis, 1954.

———. *Jubilate Agno*. Vol. 1 of *The Poetical Works of Christopher Smart*, ed. Karina Williamson. Oxford: Clarendon, 1980.

———. *Rejoice in the Lamb: A Song from Bedlam*, ed. William Force Stead. London: Jonathan Cape, 1939.

Spoel, Philippa M. "Rereading the Elocutionists: The Rhetoric of Thomas Sheridan's *A Course of Lectures on Elocution* and John Walker's *Elements of Elocution*." *Rhetorica* 19 (2001): 49–91.

Walker, Jeanne Murray. "'Jubilate Agno' as Psalm." *Studies in English Literature* 20 (1980): 449–459.

Williamson, Karina. "Introduction." In *Jubilate Agno*, vol. 1 of *The Poetical Works of Christopher Smart*, ed. Karina Williamson, xv–xxxi. Oxford: Clarendon, 1980.

———. "Surfing the Intertext: Smart Among the Moderns." In *Christopher Smart and the Enlightenment*, ed. Clement Hawes, 235–281. New York: St. Martin's Press, 1999.

Wordsworth, William. "The Tables Turned." In *Wordsworth's Poetry and Prose: Authoritative Texts, Criticism*, ed. Nicholas Halmi, 59–60. New York: Norton, 2014.

CHAPTER 4

A Volatile Unity: Coleridge, Starling Murmurations, and Romantic Form

Gavin Sourgen

<table>
<tr><td>condensing</td><td>now</td><td>coleridge</td></tr>
<tr><td>now</td><td>they</td><td>saw</td></tr>
<tr><td>glimmering</td><td>formed</td><td>starlings</td></tr>
<tr><td></td><td></td><td></td></tr>
<tr><td>a</td><td>a</td><td>in</td></tr>
<tr><td>volatile</td><td>square</td><td>vast</td></tr>
<tr><td>unity</td><td>now</td><td>flight</td></tr>
<tr><td></td><td></td><td></td></tr>
<tr><td>thickening</td><td>a</td><td>like</td></tr>
<tr><td>deepening</td><td>globe</td><td>a</td></tr>
<tr><td>blackening</td><td>now</td><td>body</td></tr>
</table>

Even the most casual reader of British romantic poetry is acquainted with its rich and pervasive ornithological import. The poetic treatment of birds and their significance to the development of a poetics of organic unity or lyrical spontaneity is prevalent in the work of all major lyricists from the latter half of the eighteenth century onward. Whether elevated as preeminent symbols of unbridled liberty, like Shelley's skylark, which ascends "Higher still and higher / . . . Like a cloud of fire" (6–8); celebrated for their displays of emancipated fervor, like the freed bird in Felicia Hemans's short lyric, which bursts forth "With the spirit that panted through heaven to soar" (31); or coveted for their melodic candor, like Keats's "light-winged Dryad of the trees," which, "In some melodious plot / Of beechen green, and shadows numberless, / Singest of summer in full-throated ease" (7–8), birds come to exemplify a romantic idealism centered on preindustrial autonomy and unconscious artistic production. In a typical response to the ubiquity of such symbols and their universal application to a naturalized aesthetics endemic to the romantic period, David Perkins argues that "birds in poetry [between 1750 and 1850] always represented 'nature' by metonymy or synecdoche"; that is to say, they held a fundamentally representational function. More than this, and as is often the case with such criticism, he emphasizes the sacred, uninfringeable state of those representations: "Of all animals a wild bird especially could represent nature as poets dearly wished nature to be—inviolable—and this is one reason why birds were the central image in so many poems" (141).

Figures of untarnished inspiration and organic sincerity they may be, but Perkins's alertness to the emblematic function of birds during these years, however pertinent, is also somewhat reductive. While he hints at an inherent instability in the relationship between signifier and signified by intimating toward a "nexus of expressive possibilities and limitations," his final emphasis on the consecrated purity of the symbols takes that representative function at face value. The birds may be endowed with privileged latitude, indivisible harmony of expression, and aesthetic immortality, but they also carry with them a more resonant dynamism and a deeper volatility, which an approach that essentializes symbols, particularly avian ones, repeatedly overlooks. It must not be forgotten that Shelley's skylark issues forth "a flood of rapture so divine" that it remains "unseen" and then "hardly see[n]," while the unearthly status of the mythological bird in Keats's illustrious ode seduces its speaker into thoughts of annihilation: to "drink, and leave the world unseen."

Very frequently in romantic poetry, such idealisms of artless freedom and lyrical ease are held in tension with the loss of authorial command and the brooding awareness—through the self-conscious rhetoric of such poems—that natural ideals are constitutionally naïve: "the project of the symbolic," as Paul de Man suggests, "haunted by the inauthenticity of its effort to identify with a temporality not properly ours" (197). In a recent study of the organic in romantic poetry, Denise Gigante tacitly endorses a resilient critical heritage by arguing that romantic poets "were all committed to defining and representing the incalculable, uncontrollable—often capricious, always ebullient—power of vitality" (3), but, as her boisterous adjectives deliberately suggest, that very same ebullience also threatens to collapse the purity of its own abundant animation. The metonymic vitality of such symbols and their dangerous inbuilt resistance to semantic constancy are never comfortably separated. As Tilottama Rajan points out in relation to Wordsworth's *The Prelude*, such poems also "deconstruct [their] own assumptions by associating nature itself with the cataclysms of historical experience . . . that make chaos and discontinuity at least as primary as organic unity" (16). Romantic poets and artists such as Samuel Taylor Coleridge and J. M. W. Turner, known for their conscious turn from conspicuously artificial forms to an organic aesthetics of internal agreement born out of the innovative spirit of the French Revolution, also carried with them a tenacious sense of tradition and an inbred skepticism about the nature of that aesthetic revolution. In their various responses to the aestheticizing of a mind in vital relation to nature, romantic artists were seldom bound to a twentieth-century critical methodology that holds construction and dissolution in starkly divergent terms. Instead, they offered more complex formal deliberations on the nature of containment and energy, on the latent and enriching quarrel, as it were, between principles of organization and primordial imperatives.

Rather than seeing the emblem of birds in flight as a flawlessly conceived aesthetic totality, it may be more productive to identify the ways in which the impulse toward a self-generated fusion of mind and nature finds itself in fecund conflict with the need for contradistinction. To that end, I turn to a well-known, though seldom examined, notebook entry of 1799 by Samuel Taylor Coleridge, in which he remarks on the curious shapeshifting effect of a cluster of starlings in flight:

> November 27th—a most interesting morning. 1799. Awoke from one of my painful Coach-Sleeps, in the Coach to London. It was a rich Orange Sky like that of a winter Evening save that the fleecy dark blue Clouds that rippled above it, shewed it to be morning—these soon became a glowing Brass Colour, brassy Fleeces, wool packs in shape/rising high up into the Sky. The Sun at length rose upon the flat Plain, like a Hill of Fire in the distance, rose wholly, & in the water that flooded part of the Flat a deep column of Light.—But as the coach went on, a Hill rose, and intercepted the Sun—and the Sun in a few minutes *rose* over it, a compleat 2nd rising, thro' other clouds and with a different Glory. Soon after this I saw Starlings in vast Flights, borne along like smoke, mist—like a body unindued with voluntary Power/—now it shaped itself into a circular area, inclined—now they formed a Square—now a Globe—now from complete orb into an Ellipse—then oblongated into a Balloon with the Car suspended, now a concave Semicircle; still expanding, or contracting, thinning or condensing, now glimmering and shivering, now thickening, deepening, blackening! (*Collected Notebooks*, 39)

This extraordinary account of contour in motion is dense with content and richly suggestive in form: a bristling correspondence of active, vitalizing forces pirouetting in orbit; an organizational drama of collective formation, change, and creative association. The advance toward a fluid totality and imminent risk of dissolution by sudden redirection—the centripetal and centrifugal—engage each other in a dynamic interaction that represents the high point of organic self-organization at the same time that it tempts vagueness with an inherent resistance to closure. Coleridge's depiction—part imaginative interchange, part descriptive awe—directs our attention less to an epitomizing symbol than to a compelling (if paradoxical) process.

The difficulty raised by Coleridge's magnificent response to the starling murmuration and relentlessly pursued in subsequent reflections (he revisited the episode on three separate occasions in his notebooks) is the simultaneous invigoration and ever-present volatility of a self-generating and self-maintaining form. As Richard Holmes, his most acclaimed biographer, says of the episode:

> It is an image of shifting energy and imagination; a protean form or a force-field, lacking fixed structure or outline; a powerful personality without a solid identity, or unified will—"without volition." Clearly, this was some sort of self-image for Coleridge, both stimulating in its sense of freedom, of "vast flights"; and menacing in its sense of threatening chaos or implosion, "thickening, deepening, blackening." (254)

While Holmes understandably focuses his interpretation on the troubled relation of unstable boundaries in nature to those of selfhood, he also seizes upon the dramatic force of the description as a potent reflection on the nature of inspiration and hazard. The principal mode of organization in Coleridge's account of the clustered fluidity of starlings in flight is neither an invisible and indivisible organic unity nor one of unencumbered process. Rather, it is the discursive interplay of an ordering perception energized by the brilliant theatricality of natural forces in spontaneous, if mysterious, coordination. It endorses Karl Kroeber's contention that, "the Romantic imagination . . . seeks to create, not simply by opposition or disjunction, but also by unification and interpenetration" (334). In their morphological patterns (biological and linguistic) the starlings fashion meaning and indeterminacy alike, and it is for this reason that neither deconstructive methodologies (which depend on differences and oppositions and consequently must annihilate continuity) nor the coalescing impulses of symbolism (which purport an idealizing spiritual fusion) present adequate paradigms for grasping the great range of associations they evoke.

In the only sustained account of the poetic significance of Coleridge's reflection, Peter Anderson sees the "indeterminate menace of the starlings' 'will to form'" as a simultaneously constructive and deconstructive rumination on the act of writing poetry: "They simultaneously allude and elude, disclose but never close, only thicken, deepen, blacken like aggregations of words on a page" (55). Anderson's analysis is intelligently conceived and elegantly articulated, striking at the heart of the pulsating contraries in the episode, but I would like to extend his claims beyond the realm of poetic composition to proffer a few broader contentions about the distinctive character of romantic form. As Gigante suggests, "even for Coleridge, on whom subsequent ideas of organic form have been based, living or organic form was never equivalent to undifferentiated unity. Instead, the unpredictable vitality of living form, its very liveliness—protean, procreative, for some terrifying—served as a model for 'genuine' art" (4). The implications of this paradoxical assertion for the reconceptualization of form is central to my inquiry.

Coleridge's Starlings and Romantic Dynamism

I wish to show, by way of a review of further key reflections of birds in other major pieces by Coleridge alongside the works of his predecessors, contemporaries, and inheritors, including Dante, Thelwall, Carlyle, William Wordsworth, Byron, and Ted Hughes, how the starling episode is fundamentally romantic in a unique yet very general way: It admits and indeed in its form asserts the simultaneously delimited and ambiguous nature of the imagination in vital connection with the phenomenological. It is, mutually, a serendipitous occurrence (a chance encounter with sublime theatricality) and a mingling of the intuitive and the aesthetically encoded that reflects both the general ideological crisis of the late eighteenth century and the expressive revolutions that accompanied it. If the verb phrase "borne along" carries with it a sense of determined motion or decisive contour, the nouns that it transports ("smoke" and "mist") are amorphously imprecise, while the curious formulation of "a body unindued with voluntary Power" is strangely assertive in its negation of Will. Coleridge's thrusting, allusive description of the starlings in active agreement gives deference to common icons and universal associations at the same time that it lends energy to the promise of dissolving them. A close connection is thus suggested between the symbolic and psychological functions of nature in active relation to the perceiving self. The sporadic use of dashes and impressionistic deferment of meaning, coupled with the exactness of the shapes evoked—the "Square," the "Globe," the "Ellipse," and the "Semicircle"—renders a compositional aesthetic that does not simply put things together but moves from one to another, knitting webs of unity and difference between them. Logocentrism exists, but it also gives way to a richer, more indeterminate verbal texture in which meaning is not simply contained by the images described but created among them.

The redolent particularity of these fluid patterns and their emblematic connotations take us far beyond the intimate significance of the episode as a private notebook entry; we are driven, instead, to a larger awareness of the formally conscious relationship between subjective and objective modes of artistic engagement. The main thrust of this account lies somewhere between a neoclassical preoccupation with universal value and a modernist syntax insistent on a "semantic flow towards meaning whose status is primarily psychological" (Watt, "Impressionism and Symbolism," 313). It hints at but never achieves the elusive nature and hazy contours of modernist atmospherics so deftly verbalized by Virginia Woolf in "Modern Fiction" when she suggests that "Life is not a series of gig lamps symmetrically arranged; life is a luminous halo, a semi-transparent envelope surrounding us from the beginning of consciousness to the end" (149). Take these lines from Dewhurst Bilsborrow's dedication to "Erasmus Darwin on His Work Entitled *Zoonomia*" (1794), which, in demonstrating "How the first embryon-fibre, sphere, or cube, / Lives in new forms,—a line,—a ring,—a tube," employ a language and tone uncannily similar to Coleridge's description of the starling murmuration:

> Now in strong lines, with bolder tints design'd,
> You sketch ideas, and portray the mind;
> Teach how fine atoms of impinging light
> To ceaseless change the visual sense excite;
> While the bright lens collects the rays, that swerve,
> And bends their focus on the moving nerve.
> How thoughts to thoughts are link'd with viewless chains,
> Tribes leading tribes, and trains pursuing trains;
> With shadowy trident how Volition guides,
> Surge after surge, his intellectual tides;
> Or, Queen of Sleep, Imagination roves
> With frantic Sorrows, or delirious Loves . . .
>
> (37–48)

Two complementary concepts come together here to conceive and foster the romantic poet's deepest triumph of form: their conviction that a poem should proceed from an elemental force flowing in and through the life of things and their attention to "strong," if "viewless," conjunctions. Just as the fluid yet purposeful motion of the starlings is likened to "a body unindued with voluntary Power," the near-allegorical figure of Volition in Bilsborrow's description of a pulsating anatomy taking shape exercises its control with the furtive maneuvering of a "shadowy trident." The movements are "ceaseless" and mighty ("Surge and surge") but also coalescent ("thoughts to thoughts," "Tribes to tribes," "trains pursuing trains"). In this we see how romantic poetry aspires to the combined effects of a compact or substantial style that compels concise clauses *and* an unceasing form of rhythmic impetus that necessitates protracted phrasing beyond its bounds. It revels in expanding and contracting motions against expansive and constrictive ones.

Romantic art is conceptual as well as concrete, technical as well as transcendental; while it yearns for an "uncompromising image of a life force or inner necessity" (Fairer, *Organising Poetry*, 22), it revels in the light and shade of dynamic contradistinction. If, for instance, Coleridge's writing seeks at times to govern its internal energies by method, claiming that the wild beauty of "Christabel" "depend[ed] for its beauty always, and often even for its metrical existence, on the *sense* and *passion*" (Coleridge, *Collected Letters*, 3:112), it delights elsewhere in the luminous haziness of uncertainty: "The elder Languages fitter for Poetry because they expressed only prominent ideas with clearness, others but darkly—Therefore the French wholly unfit for Poetry; because [all] is clear in their Language" (*Collected Notebooks*, 13). To understand better the philosophical and aesthetic significance of the starling configurations for Coleridge, and the centrality of such a depiction to a romantic poetics of dynamic coalescence, it is worth emphasizing—even exaggerating—the differences between fixed and living forms. To that end, consider a more timid reflection on bird formations by the same author. In a rumination of considerably different import, Coleridge ponders the coordinated movements of linnets in a domesticated setting:

> Observed in the garden of Eaton House the flight of the Brown Linnets, a large flock of whom I had repeatedly disturbed with my foot-fall as I walked by the thicket./1. Twinkling of wings. 2. Heavy and swanlike rise & fall, yet so that while one was rising, another was falling—& so 4. Their sweet straight onward motion/they swam on, not with speed or haste, much less hurry, but with easy natural Swiftness—& then [a] graceful wheel round one half of a circle or more, & then cut straight the diameter of it—4. Their change of position among themselves/right to left, hindw[ard] to the front, vanguard to the rear—these four motions all at once in a beautiful Whole, like a Machine. (*Collected Notebooks*, 57)

If the balance of opposing motions in the linnets' patterning—one rising, another falling—constitutes an aesthetic unity, it is of a neoclassical complexion rather than a romantic one. Where the starlings enact a potent proliferation of contrary active verbs, "expanding, or contracting, thinning or condensing," the shifting arrangements of the linnets are conspicuously more moderate: interacting with "sweet straight onward motion" and "easy natural Swiftness." While lacking the abrupt (although admittedly short-lived) rupture by an unnamed disturbance in which "All suddenly mount/And scatter wheeling in great broken rings/Upon their clamorous wings" (10–12), the picturesque permanence of this impression is reminiscent of Yeats's "Wild Swans at Coole," whose tableau of social agreement is tinged with the sadness of mortality, consciously framed. The linnets trade stations with military exactness, leading to an image of their final achievement that is couched in fundamentally inorganic terms: "a beautiful Whole, like a Machine." The loveliness of modestly active parts conforming to an automated whole that Coleridge observes here is charming and speaks of a deeper fascination with meticulous order than has generally been acknowledged in his thought and verse, but it also betrays his most comprehensive articulation of organic form, in a passage translated from Friedrich Schlegel delivered on several lectures on the theme: "The form is mechanic when on any given material we impress a pre-determined form . . . the organic form of the other hand, is innate; it shapes as it develops itself from within, and the fullness of its development is one and the same with the perfection of its outward form. Such is life, such is the form" (*Collected Works*, 5.1:495). The linnets produce a harmony, but it is indubitably of a technical kind: one that effects pleasant consideration without activating the sublime imagination, and one enjoyed from a quiet distance without demanding vigorous engagement.

By contrast, the expanding, swelling configurations of the starlings in mesmeric coordination, surging as they do with the elementary energies of innate matter, impel the observer (and reader) to inhabit the active motion of their fluid clusters, to participate in the lively networks of their dramatic associations. Marking no point of origin and no terminal station, the murmuration affects a heterogeneous give-and-take arrangement that is nonetheless sinuously continuous. Different shapes succeed and grow out of one another to form not a picture but a process: an intangible living power most memorably expressed in lines from "Tintern Abbey" as a "A motion and a spirit, that impels/All thinking things, all objects of all thought,/And rolls through all things" (101–103). Organic theory, especially of the kind that flourishes in the final decades of the eighteenth century, pivots around this move from dead fixities to dynamic vivacities, emphasizing not static agreement but purposeful activity. For Frederick

Engels, this investment in process is the fundamental marker of modern consciousness:

> The great basic thought that the world is not to be comprehended as a complex of readymade *things*, but as a complex of *processes*, in which the things apparently stable no less than their mind images in our heads, the concepts, go through an uninterrupted change of coming into being and passing away, in which, in spite of all seeming accidentally and of all temporary retrogression, a progressive development asserts itself in the end—this great fundamental thought has, especially since the time of Hegel, so thoroughly permeated ordinary consciousness that in this generality it is now scarcely ever contradicted. (Marx and Engels, *Selected Works*, 609)

Although Engels makes reference to Hegel when he marks an irrevocable turn from a preconceived to a fluctuating sensibility, he has, of course, Hume's claim in *A Treatise of Human Nature* in mind—that the self is little more than "a bundle or collection of different perceptions, which succeed each other with an inconceivable rapidity, and are in perpetual flux and movement" (239).

As both reflections intimate, this transition from the programmed to the abundantly active was hardly a straightforward one, but the distinction between a constrictive and generative form—between notions of form as a noun and form as a verb as it were—was crucial to romantic poets like Coleridge. He insisted on it in no uncertain terms in his *Biographia Literaria*: "Remember that there is a difference between form as proceeding, and shape as superinduced;—the latter is either the death or imprisonment of the thing;—the former is its self-witnessing and self-effected sphere of agency," and however much he struggled to maintain a clear boundary between them in his poetic treatment of form, Coleridge was to dwell on those conflicting conceptions throughout his life. They are most despondently conveyed in a letter to Godwin in 1802 that seizes upon the damaging implications of being severed from the idealistic golden thread of living matter:

> [P]artly from an unhealthy & reverie-like vividness of Thoughts, & (pardon the pedantry of the phrase) a diminished Impressibility from Things my ideas, wishes, & feelings are to a diseased degree disconnected from motion & action. In plain & natural English, I am a dreaming & therefore an indolent man I am a Starling self-incaged, & always in the Moult, & my whole Note is, Tomorrow, & tomorrow, & tomorrow. (*Collected Letters*, 2:782)

Whether by an overactive or underactive imagination, Coleridge, in typically self-flagellating fashion, finds himself devoid of the "high spiritual instinct" that "impel[s] us to seek unity" (*Biographia* 2:72). By figuring his despondency in the image of a shoddy, self-imprisoned starling reduced to the mimetic repetition of a single statement of endless deferral, he reiterates the importance of the species as both a personal and universal symbol for enlivening mobility.[1] The idea lingers in a more general manner over a note recorded in London only two days after his transcendent encounter with the dazzling formations on a coach from Somerset: "The immoveableness of all Things thro' which so many men were moving—harsh contrast compared with the universal motion, the harmonious System of Motions, in the country & every where in Nature.—In this dim Light London appeared to me as a huge place of Sepulchres thro' which Hosts of Spirits were gliding" (*Collected Notebooks*, 39). While the actual mechanics that inform this discrepancy between an essential kinesis and a vague, disembodied fluidity of the kind portrayed in later accounts of urban consciousness by Dickens, Conrad, and T. S. Eliot is never elucidated, its enduring implication for a romantic poetics of vitality is clear.

Parts and Wholes

On a philosophical level, the starlings represent the perfect relationship between parts and wholes, that is, the magical consolidation of discrete elements in a form that is unified but not mechanically bound to static delineation, or, to summon a definition from John Thelwall's 1793 "Essay towards a Definition of Animal Vitality," "The stimuli necessary for the production and sustainment of Life . . . absorbed and properly diffused through [an] organized frame" (20). The opening lines of a recent scientific study on the enigmatic nature of this process give credence to this mystery: "Self-organization and the spontaneous emergence of order in biological systems does [*sic*] not come much more spectacular than in large flocks of starlings. At dusk, huge flocks move above the roost, exhibiting beautiful collective patterns. There is no leader in the group, and the collective movement is a unique consequence of local interactions between individuals" (Cavagna et al., "Diffusion of Individual Birds," 1). In this way, they fit Coleridge's philosophy of organic form quite comfortably. The surreptitious acquiescence of rival elements that embody his deepest convictions about structural achievement is situated in a global order that emerges out of spontaneous self-organization even if it is emphatically unclear in its terms and conditions. Yet, the whole is also larger than the sum of its parts, and as Seamus Perry points out in a passing deliberation on the episode, "The starlings are each individual and autonomous, and yet part of the greater, changing life of the whole flock, an apprehension at once of the discrete Many and of the encompassing One." (30) In aesthetic contemplation of the flock's formation, Coleridge becomes aware of a sublime totality demonstrating qualities not conspicuously related to the properties of, and interactions between, its constituent parts. Implicit in the constitutive binaries of difference is an urge toward purposeful action in which all elements of an organism are reciprocally dependent and reciprocally formative: "the body unindued with voluntary Power." As Kant explains in his "Critique of Teleological Judgment":

> Just as each part exists only as a result of all the rest, so we also think of each part as existing *for the sake of* the others and of the whole . . . [instead] we must think of each part as an organ that *produces* the other parts (so that each reciprocally produces the other). . . . Only if a product meets that condition [as well], and only because of this, will it be both an *organized* and a *self-organizing* being, which therefore can be called a *natural purpose*. (*Critique of Judgment*, 253)

There are, however, some fundamental difficulties that arise when converting such a symbol of spontaneous coadunation from a philosophical ideal to an artistic principle. For one thing, it calls into question the nature and magnitude of authorial command. In a celebrated articulation of the poet's capacity to bind contraries with imperceptible accord through the power of the imagination, Coleridge gestures toward a reconciliation of the competing notions of form as embodied essence and form as material process that nevertheless reveals a deep anxiety over its ability to do so:

> The poet, described in ideal perfection, brings the whole soul of man into activity, with the subordination of its faculties to each other, according to their relative worth and dignity. He diffuses a tone, and spirit of unity, that blends, and (as it were) fuses, each into each, by that synthetic and magical power, to which we have exclusively, appropriated the name of imagination. . . . This power, first put into action by the will and understanding and retained under their irremissive, though gentle and unnoticed control . . . reveals itself in the balance or reconciliation of opposite or discordant qualities: of sameness with difference; of the general, with the concrete; the idea, with the image; the individual, with the representative; the sense of novelty and freshness, with old and familiar objects; a more than usual state of emotion, with more than usual order; judgment ever awake and steady

> self-possession, with enthusiasm and feeling profound or vehement; and while it blends and harmonizes the natural and the artificial, still subordinates art to nature; the manner to the matter; and our admiration of the poet to our sympathy with the poetry. (*Biographia*, 2:15–16)

As remarkable as this passage is, ebbing and flowing in the beauty of an undulating logic, it is also riddled with the imperfect suppression of unwanted inconsistencies: the unqualified evaluative hierarchy, the substitution of method with abstract qualities, the unexplained relation of the will (which initiates before fading) to "a more than usual state of emotion."[2] Embedded in such deliberations is an uncomfortable awareness of a possible inauthenticity in the method of organization, in the necessity of the will to set things in motion and to maintain a "gentle and unnoticed control." As Kant himself had conceded, "the order and regularity in the appearances, which we entitle *nature*, we ourselves introduce. We could never find them in appearances, had not we ourselves, or the nature of our mind, originally set them there" (*Critique of Pure Reason*, 147).

The obvious dilemma growing out of this awareness is the ever-present possibility of the ideal compromising, and ultimately superseding, the natural. Not art subordinated to nature but nature subordinated to art. As Coleridge says elsewhere, with an equally uncomfortable dualism hovering over his claims:

> The spirit of poetry, like all other living powers, must of necessity circumscribe itself by rules, were it only to unite power with beauty. It must embody in order to reveal itself; but a living body is of necessity an organized one—and what is organization but the connection of parts to a whole, so that each part is at once end and means. (Coleridge, *Collected Works*, 5.1:496)

The delicate high-wire act of seamless reconciliation he wishes to perform here straddles the desire to see nature fulfilling an ideal or universal totality and a wish for it to remain vitally active. This sentiment is memorably rearticulated by Shelley, who declares that "Poetry turns all things to loveliness; it exalts the beauty of that which is most beautiful, and it adds beauty to that which is most deformed; it marries exultation and horror, grief and pleasure, eternity and change; it subdues to union under its light yoke all irreconcilable things" (41). This takes us deeper into the problematic nature of what I have been calling the "dynamic coalescence" of romantic aesthetics, whereby the need to "have things both ways,"[3] while richly productive, also means that one might not have it either way. In this instance, the unifying necessity of discrete entities—helped along by the imperial faculty of the poet—threatens to be reduced from a spirited arrangement of animate powers to a mechanical assemblage of a neoclassical kind put forward by Burke: "We compensate, we reconcile, we balance. We are enabled to unite into a constituent whole the various anomalies and contending principles that are found in the minds and affairs of men. From hence arises, not an excellence in simplicity, but one far superior, an excellence in composition" (249). The murmuration that Coleridge witnesses certainly has a life of its own, but that gives way, at least momentarily, to an unwelcome structural associationism in the author, who sees the mobile cluster conforming to preexisting forms, the "Square," the "Globe," moving "from complete orb into an Ellipse—then oblongated into a Balloon with the Car"—and hinging dangerously close to the poet's own pejorative characterization of the "FANCY," which "must receive all its materials ready made from the law of association" (*Biographia* 1:103–104).

By the same token, the potentially overactive and irrepressible energy of the starlings—embodied in the rapid exchange of verbs, "expanding . . . contracting . . . thinning . . . condensing . . . glimmering . . . shivering . . . thickening . . . deepening . . . blackening"—imperils a complete lack of authorial control and a fundamental aesthetic disunity. The murmuration enacts an impressively fluid whole, but as researchers point out, "imitation and mutual alignment are never complete, there is always an amount of uncertainty or arbitrariness in the individual choices. As a consequence, flight directions between neighbours are very similar, but not identical, differing by small 'random' fluctuations" (Cavagna et al., "Diffusion of Individual Birds," 2). The extent to which the flock resists static totality by an imperceptible but essential divergence between its individual flecks of agency is expressed in its seemingly infinite capacity for reorganization. A striking engagement with this mode of intrinsic organization through unbounded internal difference can be seen in Thomas Carlyle's enthusiastic description of the French mob:

> Other mobs are dull masses; which roll onwards with a dull fierce tenacity, a dull fierce heat, but emit no light-flashes of genius as they go. The French mob, again, is among the liveliest phenomena of our world. So rapid, audacious; so clear-sighted, inventive, prompt to seize the moment; instinct with life to its finger-ends! . . . Your mob is a genuine outburst of Nature; issuing from, or communicating with, the deepest deep of Nature. (1:216)

Carlyle's language, like that of Coleridge, is positively energized, but however much he may revel in the inbred spontaneity of a large, coordinated body of vibrant particulars (his judgment clouded somewhat by historical and national distance), on this occasion the pulsating collection of individualities contains the dormant seeds of arbitrary or diabolical power, power that is, as first defined by Johnson, inherently volatile in its throbbing extensiveness: "1. Despotick; absolute; bound by no law; following the will without restraint . . . 2. Depending on no rule; capricious" (in Keach, *Arbitrary Power*, 3). As an image of artistic creation, the starlings, like the mob, raise problems concerning the proper relation of freedom to entropy, for as much as Coleridge praises the unconsciously willful efforts of the secondary imagination, which "dissolves, diffuses, dissipates, in order to re-create," it also underlines how seamless accretion is vulnerable to dissolution as shapes are repeatedly disintegrated in order to be reformed and reconstituted.

And if the collective risks imminent disbandment by its particles, it also threatens to grow into an ominous magnitude several times larger than its individual parts. Kant has observed of the colossally grotesque: "An object is monstrous, if by its magnitude it nullifies the purpose that constitutes its concept" (*Critique of Judgment*, 109). While the principal function of a starling murmuration has never been made clear (arguments that it is a form of sublime theatricality designed to ward off predators remain unconvincing), the lack of distinction, or contradistinction, in the giant swirling vortex "thickening, deepening, blackening" is apocalyptic in scope. A similar image of portentous blanketing is on display in Wordsworth's description of an immeasurable grotto in *The Prelude*:

> Substance and shadow, light and darkness, all
> Commingled, making up a canopy
> Of shapes, and forms, and tendencies to shape,
> That shift and vanish, change and interchange
> Like spectres—ferment quiet and sublime.
>
> (VIII, 719–723)

Caught in what appears to be an overwhelmingly active correspondence between the elements necessary for perception, Wordsworth's traveler finds himself disoriented by a space that is notable for its magnitude but also heinously ghostly and provisional. Byron employed parallel terms in his response to Rubens, whose paintings he described to Hobhouse as "not nature . . . not art . . . [but] an assemblage of florid night-mares" (5:73), and again in the opening lines of *Hints from Horace*, where he contended that the integrity of classical proportion had been so thoroughly abandoned by certain artists that they fashioned creations "sillier than a sick man's dreams,/ . . . a crowd of figures incomplete,/Poetic night-mares, without head or feet" (12–14).[4] The sheer excess of signifiers and almost unmappable changes of direction in the starling murmurations support Anne K. Mellor's investment in the tenets of English romantic irony, which she defines as "a conception of the universe as random motion, as a fertile chaos that always throws up new forms" (4), but they also represent something darker, a copious force of doom. This treacherous conception of unstable collective agreement dominates a poem by Ted Hughes (a prominent figure in that ever-ending line of "Last Romantics") entitled "Starlings Have Come":

A horde out of Sub-Arctic Asia
Darkening nightfall, a faint sky-roar
Of pressure on the ear.

More thicken the vortex, gloomier.

A bacteria cyclone, a writhing of imps
Issuing from a hole in the horizon
Topples and blackens a whole farm.

Now a close-up seething of fleas.
And now a silence—
The doom-panic mob listens, for a second.
Then, with a soft boom, they wrap you
Into their mind-warp, assembling a nightmare sky-wheel
Of escape—a Niagara
Of upward rumbling wings—that collapses again

In an unmanageable weight
Of neurotic atoms . . .

(609)

That they are harbingers of oblivion is evident in the title of the poem, which intimates toward the unwelcome arrival of a plague. This feeling of ominous intrusion is furthered by descriptions of the flock as a "horde," a "bacteria cyclone, a writhing of imps," and a "seething of fleas" hanging darkly in the sky. The birds themselves carry a mythological baggage (as John Aubrey says of the starlings at Stonehenge, "The Welsh doe call Stars *Sturni Adar y Drudwy*, i.e. Aves Druidum, 'birds of the druids'"), but this depiction moves beyond omens; it is in many ways the Kantian grotesque realized, "an unmanageable weight/Of neurotic atoms" on the verge of a dangerous copiousness. Although Hughes's description shares many of the verb roots that Coleridge employed ("darkening," "thicken," "blacken"), there are fundamental differences between the latter's account of dynamic abundance and the image of gloomy immanence portrayed by Hughes. The internalization of the connections in Coleridge's piece and the excited but fitful way in which they move between states and from one shape to another emphasize wonder rather than portent. The movement is one of inexorable ongoing dilemma that is, nonetheless, vigorously generative.

Coleridge's misgivings about the general effect of sublime particulars disconnected from a broader aesthetic unity are registered in his criticism of Gibbon's *Decline and Fall*, which he saw a string of detached anecdotes: "When I read a chapter in Gibbon I seem to be looking through a luminous haze or fog, figures come and go, I know not how or why, all larger than life, or distorted or discolored; noting is real, vivid, true; all is scenical, and as it were, exhibited by candlelight" (*Table Talk*, 245). It is curious that Gibbon's masterpiece, so frequently referenced as an exemplum of authorial command in the eighteenth century, would have troubled Coleridge so and that his criticism would be directed at, of all places, its formlessness. But for the romantic poet who seeks not simply abiding but constitutive unity, Gibbon's collection of shimmering fragments is a form of incoherence, a product of the Enlightenment dualism that "replaced a providential, vital, and companionable world by a world of particles in purposeless movement" (Bloom, *Romanticism and Consciousness*, 217). To understand how deeply this diverges from a modernist attraction to shifting impressionism, we need only to juxtapose it to the celebration of tonal abstractions in Virginia Woolf's response to *Decline and Fall*:

> The innumerable figures are suffused in the equal blue of the far distance. They rise and fall and pass away without exciting our pity or our anger. But if the figures are small, they are innumerable; if the scene is dim it is vast. Armies wheel; hordes of barbarians are destroyed; forests are huge and dark; processions are splendid; altars rise and fall; one dynasty succeeds another. The richness, the variety of the scene absorb us. (*The Death of the Moth, and Other Essays*, 82)

The starlings Coleridge encounters are "vast and innumerable"; they too "rise and fall," but their conversion into an aesthetic totality is navigated by the poet's binding gaze:

> My Spirit with a fixed yet leisurely gaze
> Following its even yet quietly
> Changing Clusters of Thoughts,

As the outward Eye of a happy
Traveller a flock of Starlings . . .

(*Collected Notebooks*, 52)

This is not to say that a romantic poetics of interminable motion does not exist or that it is not everywhere apparent in their work. Countless personal, philosophical, and poetic reflections demonstrate an opposition to inert forms of art, especially ones that are too confident in their resolutions. Take, for instance, another notebook fragment by Coleridge that reflects his negative view of picturesque verse, whose surging momentum is marred by superfluous pauses: "Dr. Darwin's Poetry, a succession of Landscapes or Paintings—it arrests the attention too often, and so prevents the rapidity necessary to pathos—it makes the great little" (3). As Thomas McFarlane points out, in the years succeeding the French Revolution, "A new fascination with electrical dynamism permeated metaphorical conceptions: 'It is impossible to read the compositions of the most celebrated writers of the present day', said Shelley in 1821, 'without being startled with the electric life which burns within their words'" (7). Considerations such as these are often employed to vindicate sweeping critical assertions about a romantic investment in the world of infinite activity and an accompanying poetic syntax that performs "the process of its own becoming" (Armstrong, "'Tintern Abbey,'" 21). Whether by insisting on a seamlessly continuous motion, like Abrams and Armstrong, or an abundantly chaotic interchange, like Mellor, the general urge toward an appreciation for unstoppable fluidity remains a defining feature of romantic scholarship.

Unity and Force of Expression

The problem with such a pervasive critical acceptance is that it fails to see the romantic poet's unique facility for finding a form and language that rises to the demands of being concurrently lucid and suggestive, one that produces a powerful sense of prolongation without being weakly disseminated across the page and without resorting to the systematic regularity of its neoclassical antecedents. To that end, Coleridge's account of the starlings—largely articulated through present-participle verbs—and its most famous literary precursor are much to the point:

The stormy blast of Hell
With restless fury drives the spirits on,
Whirl'd round and dash'd amain with sore annoy.
When they arrive before the ruinous sweep,

There shrieks are heard, there lamentations, moans,
And blasphemies 'gainst the good Power in Heaven.
I understood, that to this torment sad
The carnal sinners are condemn'd, in whom
Reason by lust is sway'd. As, in large troops
And multitudinous, when winter reigns,
The starlings on their wings are borne abroad;
So bears the tyrannous gust those evil souls.
On this side and on that, above, below,
It drives them: hope of rest to solace them
Is none, nor e'en of milder pang.

(5:32–46)

While these eminent lines hover with spectral attendance over Coleridge's private chronicle, they represent a considerably different artistic enterprise. The incessantly coerced, restlessly unfolding, and spatially disaggregated vision of Dante's *Inferno* speaks of a merciless supernatural power that is as tumultuous as it is relentless. Just as the "multitudinous" flock of starlings is both united as a collective of discrete elements and "borne" uncontrollably abroad by an elemental force, so too is the motley assemblage of damned souls ceaselessly swept along "On this side and on that, above, below" in an all-encompassing gust of perpetual agony. These lines exemplify the perceptual turbulence generated by rapid transitions with no end in sight, the same interminable motions that Charles Lamb saw in Hogarth's "Gin Lane" print, which left him "reeling from the effect of that diabolical spirit of phrenzy which goes forth over the whole composition" (Blunden, *Charles Lamb*, 106).

A corresponding (though subtly deviating) grammatical insistence of unremitting motion is evident in the rhetorical patterning of Coleridge's account of the murmuration borne along by perpetual change: "Now it shaped itself into a circular area, inclined—now they formed a Square—now a Globe—now from complete orb into an Ellipse." The repetitive swiftness of the construction "now . . . now . . . now" is as much a form of linguistic fluidity as it is a reflection of the precipitously shifting image before him. Tenor and vehicle are exchanged with such rapidity as to be vitally unstable, transition proliferates transition, and the speaker seems to be discarding his impressions in the very act of uttering them. This same pattern is manifest in Shelley's rumination on "Mont Blanc," where the mind and matter are engaged in incessant interchange:

> The everlasting universe of things
> Flows through the mind, and rolls its rapid waves,
> Now dark—now glittering—now reflecting gloom—
> Now lending splendour, where from secret springs
> The source of human thought its tribute brings
> Of waters—with a sound but half its own . . .
>
> (1–6)

The experiential fabric of these lines reflects a power that is concentrated but not fixed. As the speaker's impressions turn sharply upon his sensory interplay with the natural world, he enacts a desire to marry the sublime disjunction of subject and object to a Wordsworthian correspondence with the universe. The supple eloquence of the passage epitomizes the romantic poet's yearning for a connective power that is ubiquitous and innate but that does not dissolve its connections in dissipating metrical arrangements. It is reminiscent of one of the more remarkable commentaries on a Turner painting by Ruskin, one that could be read in direct relation to Coleridge's vision of starlings:

> Along this fiery path and valley, the tossing waves by which the swell of the sea is restlessly divided, lift themselves in dark, indefinite, fantastic forms, each casting a faint and ghastly shadow behind it along the illuminated foam. They do not arise everywhere, but three or four together in wild groups, fitfully and furiously, as the under strength of the swell compels or permits them; leaving between them treacherous spaces of level and whirling water, now lighted with green and lamp-like fire, now flashing back the gold of the declining sun, now fearfully dyed from above with the undistinguishable images of the burning clouds, which fall upon them in flakes of crimson and scarlet, and give to the reckless waves the added motion of their own fiery flying. (189)

Ruskin marvels at the way Turner renders nature's capacity for generating dramatically evanescent forms that emerge to disappear and in disappearing create abundant new forms. But however firmly Ruskin forges his impression on "dark, indefinite, fantastic forms" that arise as the "swell compels or permits them," we cannot say, as Matthew Arnold did of Wordsworth, that "Nature seemed to take the pen out of his hand and to write for him with her own bare, sheer, penetrating power" (xxii), because Turner's attentiveness to the connotations of shape and contour is "fitfully and furiously" apparent. The mechanics of this exchange are neither wholly natural, nor are they explicitly conceited. They testify to the romantic artist's ability to capture the fluctuations of a mind in vigorous interaction with nature and to render them in complex depth.

In a *Blackwoods* review of Byron's *Manfred*, John Wilson strikes at the heart of what I have been trying to put forward here:

> His description of the stormy night among the Alps—of the blending—the mingling—the fusion of his own soul, with the raging elements around him . . . shows that he might enlarge the limits of human consciousness regarding the operations of matter upon mind, as widely as he has enlarged them regarding the operations of the mind upon itself. (Reiman, *The Romantics Reviewed*, B.1.118)

For Wilson, the energy and dynamism of Byron's descriptions of the raging elements acting upon Manfred as he struggles to summon a corresponding power from within are an "enlarging" or expanding of consciousness almost precisely because they are "borne along" by a purposeful syntax. The romantic imagination is active and formidable, but in forsaking the abstract it denies a systematic means of expressing a settlement between formalism and spontaneity, between "the generic bias of the neoclassical poet and the psychological bias of the sublime poet" (Cronin, *Shelley's Poetic Thoughts*, 31). The resulting aesthetic is one of unity in powerful motion, the desire for a feeling of restraint and firmness in art that is not inflexible married to a general contempt for motionless sterility, for as Michael O'Neill justly points out, "the wild or transgressive energies of Romanticism frequently house themselves within literary forms of some sophistication" ("Romantic Forms," 275).

For all his mistrust of closed systems and predetermined positions, there remained an urging desire in Coleridge to reconcile those hallmarks of Augustan taste—definitive artistic contour and evaluative permanence—with internal form-developing process. Time and again in his notebook entries and correspondence, fascination is directed to the curious coexistence of boundless energies and enduring forms, as in this astonishing rumination on a waterfall: "What a sight it is to look down on such a Cataract!—the wheels, that circumvolve in it—the leaping up & plunging forward of that infinity of Pearls & Glass Bulbs—the continual *change* of the *Matter*, the perpetual *Sameness* of the Form—it is an awful Image & Shadow of God & the World" (*Collected Letters*, 2:456–457). Coleridge's engagement with the murmuration verges on the timeless and transcendental, but it is held back by the poet's attentiveness to the immediacy of time and space, to conjunctions, transitions, transformation, and to a motion that is jarring as much as it is fluid. Pivoting between the formless terror of a Burkean sublime (a mass of inchoate material) and a Popean conviction in firm cosmic arrangements, the bold inflections of the starlings are perfectly constitutive of the dazzling heterogeneity of a romantic form that is paradoxically "still expanding."

NOTES

1. The dread of a starling circumscribed is reflected in a poem from Wordsworth's juvenilia entitled "The Death of a Starling":

> Pity mourns in plaintive tone
> The lovely Starling dead and gone;
> Weep ye loves, and Venus, weep
> The lovely Starling fall'n asleep!
> Venus see, with tearful eyes,
> In her lap the Starling lies,
> While the loves all in a ring
> Softly stroke the stiffen'd wing.

2. The study of starling-flight modeling by Cavagna et al. goes some way in explaining "neighbour reshuffling," but a question still hovers over whether the border "is ruled simply by diffusion or whether there is some extra dynamical ingredient ruling the way birds remain on the border" (6).

3. As Perry so eloquently asserts, "Following a path which MacNiece, the determined pluralist, does not take, Coleridge's thinking habitually seeks to correct into oneness the apparently incorrigible plurality continually re-discovered in the sharpness of his senses; while (to look at the predicament from the other end), in the teeth of his commitment to universality and oneness, diversity and particularity continue to exert their interest—so that, in practice, the unity which he proclaims so vociferously is typically submerged by the protracted exhibition of the contradictory elements he is meant to be bringing together. This is not just an oblique way of embracing heterogeneity after all; it is trying to have things both ways" (22).

4. Byron is undoubtedly thinking about Plato, who in his *Phaedrus* argued that "Any discourse ought to be constructed like a living creature, with its own body, as it were; it must not lack either head or feet; it must have a middle and extremities so composed as to suit each other and the whole work."

WORKS CITED

Abrams, M. H. "Structure and Style in the Greater Romantic Lyric." In *Romanticism and Consciousness*, ed. Harold Bloom, 90–119. New York: Norton, 1970.

Armstrong, Isobel. "'Tintern Abbey': From Augustan to Romantic." In *Augustan Worlds*, ed. J. C. Hilson et al., 261–280. Leicester: Leicester University Press, 1978.
Arnold, Matthew. *Poems of William Wordsworth*. London: Macmillan, 1879.
Bloom, Harold, ed. *Romanticism and Consciousness*. New York: Norton, 1970.
Blunden, Edmund. *Charles Lamb and His Contemporaries*. Cambridge: Cambridge University Press, 1933.
Burke, Edmund. *Reflections on the Revolution in France*. London, 1790.
Byron, Lord George Gordon. *Byron's Letters and Journals*. Ed. Leslie A. Marchand. 13 vols. London: John Murray, 1973–1994.
Carlyle, Thomas. *The French Revolution: A History*. London: Chapman & Hall, 1837.
Cary, Henry Francis. *The Vision, or, Hell, Purgatory and Paradise of Dante Alighieri*. London: Smith, 1844.
Cavagna, A., S. M. Duarte Queirós, I. Giardina, et al. "Diffusion of Individual Birds in Starling Flocks." *Proceedings of the Royal Society B: Biological Sciences* 280, no. 1756 (2013), doi:10.1098/rspb.2012.2484.
Coleridge, Samuel Taylor. *Collected Letters of Samuel Taylor Coleridge*. Ed. Earl Leslie Griggs. 6 vols. Oxford: Clarendon, 1956–1971.
———. *The Collected Works of Samuel Taylor Coleridge*. Bollingen Series 75. Princeton, N.J.: Princeton University Press, 1969–2002.
———. *Coleridge's Notebooks: A Selection*. Ed. Seamus Perry. Oxford: Oxford University Press, 2002.
———. *Table Talk Recorded by Henry Nelson Coleridge*. Ed. Carl Woodring. Princeton, N.J.: Princeton University Press, 1990.
Cronin, Richard. *Shelley's Poetic Thoughts*. Basingstoke: Macmillan, 1981.
Darwin, Erasmus. *Zoonomia; or, The Laws of Organic Life*. 2 vols. London, 1794.
De Man, Paul. *Blindness and Insight: Essays in the Rhetoric of Contemporary Criticism*. London: Routledge, 1971.
Fairer, David. *Organising Poetry: The Coleridge Circle, 1790–1798*. Oxford: Oxford University Press, 2009.
Gigante, Denise. *Life: Organic Form and Romanticism*. New Haven, Conn.: Yale University Press, 2009.
Holmes, Richard. *Coleridge: Early Visions, 1772–1804*. New York: Pantheon, 1989.
Hughes, Ted. *Collected Poems*. Ed. Paul Keegan. New York: Farrar, Straus and Giroux, 2003.
Hume, David. *A Treatise of Human Nature*. Ed. A. D. Lindsay. London: J. M. Dent, 1934.
Kant, Immanuel. *A Critique of Judgment*. Ed. Werner S. Pluhar. Indianapolis, Ind.: Hackett, 1987.

———. *A Critique of Pure Reason*. Ed. Norman Kemp Smith. London: Macmillan, 1933.
Keach, William. *Arbitrary Power: Romanticism, Language, Politics*. Princeton, N.J.: Princeton University Press, 2004.
Kroeber, Karl. "Experience as History: Shelley's Venice, Turner's Carthage." *ELH* 41, no. 3 (Autumn 1974): 321–339.
Marx, Karl, and Frederick Engels. *Marx and Engels: Selected Works*. London: Lawrence and Wishart, 1968.
McFarland, Thomas. *Romantic Cruxes*. New York: Clarendon, 1987.
Mellor, Anne K., *English Romantic Irony*. Cambridge, Mass.: Harvard University Press, 1980.
O'Neill, Michael. "Romantic Forms: An Introduction." In *Romanticism: An Oxford Guide*, ed. Nicholas Roe. Oxford: Oxford University Press, 2005.
Perkins, David. *Romanticism and Animal Rights*. Cambridge: Cambridge University Press, 2003.
Rajan, Tilottama. *Dark Interpreter: The Discourse of Romanticism*. Ithaca, N.Y.: Cornell University Press, 1980.
Reiman, Donald H., ed. *The Romantics Reviewed: Contemporary Reviews of British Romantic Writers*. Part B: *Byron and Regency Society Poets*. 5 vols. New York: Garland, 1972.
Ruskin, John. *Modern Painters*. Abridged ed. Ed. David Barrie. London: Pilkington, 1873.
Shelley, Percy Bysshe. *Shelley's Poetry and Prose*. Ed. Donald H. Reiman and Neil Fraistat. New York: Norton, 2002.
Thelwall, John. *An Essay towards a Definition of Animal Vitality*. London, 1793.
Watt, Ian. "Impressionism and Symbolism in *Heart of Darkness*." In *Heart of Darkness*, by Joseph Conrad, 4th Norton Critical ed., ed. Paul B. Armstrong, 349–364. New York: Norton, 2006.
Woolf, Virginia. "Modern Fiction." In *The Common Reader*, 146–154. London, 1938.
Woolf, Virginia. "The Historian and 'The Gibbon.'" In *The Death of the Moth, and Other Essays*. London: Harcourt Brace and Co.,1942.

CHAPTER 5

Words Are for the Birds: "Non-reasoning Creatures Capable of Speech" in the Writings of Schreber and Poe

Joe Conway

<table>
<tr><td>a
voice
and</td><td>the
raven's
nevermore</td><td>a
phony
logic</td></tr>
<tr><td>nothing
more
than</td><td>the
nonreasoning
creature</td><td>is
what
connects</td></tr>
<tr><td>miraculously
created
birds</td><td>capable
of
speech</td><td>poe
and
schreber</td></tr>
</table>

The Xerox Machine of Modernity

In his study of Ludwig Wittgenstein, Daniel Paul Schreber, and the uncanny structural resemblance between Western philosophy and clinical schizophrenia, Louis Sass recounts the case of a patient who compared himself to a Xerox machine (66). More specifically, this man felt that the world was a series of images printed on transparent paper and that he was the copier (re)producing them. The articulation of such fantasies, in which one experiences subjectivity as a field of action limited to mechanically (re) producing an always already secondhand world, is common, Sass writes, to the literature of schizophrenia. Thus, for example, in *Memoirs of My Nervous Illness* (1903) Schreber calls himself the last real person in a world of "miraculously created puppets" (18), even while dismissing his own ideas as "falsified nonsense" (126) implanted by external voices. Such fear of facsimilitude also seems to have afflicted Schreber's most famous interpreter, Sigmund Freud. Eric Santner writes that Freud's fascination with the *Memoirs*, from which he developed his theory of paranoia, derived in part because Schreber's elaborate, systematic refutation of empirical experience resembled a psychoanalytic procedure. If the hermeneutic of the analyst merely reproduces the paranoid logic of the analysand, perhaps both "might only be repeating, might only be parroting back thoughts, words, and phrases originating elsewhere" (Santner, *Germany*, 21). Thus for Santner, Schreber's *Memoirs* and its many fantasies of verbal persecution designed to produce "compulsive thinking" (Schreber, *Memoirs*, 55) make it a key text for identifying a larger set of anxieties concerning the role of automatized language in the construction of the modern subject.

One of the most famous of the persecution fantasies described by Schreber is that of the "miraculously created birds" (194) whose capacity to speak he calls "so marvelous and like a fairy tale" (196). More correctly stated, the birds emit voice (*phone*) rather than speak words (*logos*), tirelessly repeating "senseless phrases learnt by rote," which they inject directly as "the poison of corpses" (190) into the prostrate Schreber's nervous system. Schreber declares that the talking birds are sent to "either kill me or destroy my reason" (190) by the twin deities who afflict him throughout the *Memoirs*, the "upper God" Ormuzd and the "lower God" Ariman (30). These chattering birds, as we will see, have variously been interpreted as young girls, the Holy Spirit, a magician's doves, and Kafkaesque bureaucrats. Yet while most have focused on them as signs confirming Schreber's abjection, I would like to emphasize the strategies he uses to contest the verbal toxicity they spread. In particular, I want to consider how Schreber's deployment of poetry and poetic devices—most importantly the homophonic trope of punning—momentarily disrupts the mechanisms of linguistic domination that serve the "rotten" (186) symbolic regime he calls, throughout his *Memoirs*, "The Order of the World."

Speaking for the general critical consensus, Lacan rightfully points out that Schreber's *Memoirs* stress "the predominance of the function of the signifier" (176). Nowhere is this more evident than in "the phenomenon of the talking birds" (Schreber, *Memoirs*, 196) in which language is removed from any semantic dimension and instead experienced by Schreber as so much vibrating matter entering his flesh. Following Santner's lead, I will explore how these birds and what they voice figure as a trope for automatized language. Yet rather than treat the birds only in terms of a semantic lack, I want to emphasize the sonorous excess that they, in collaboration with Schreber, help produce in the *Memoirs*. In short, I want to focus on how Schreber's strategy of punning as a means of resisting the talking birds' power participates in a poetic program that advocates for restoring the primacy of *phone* over *logos*, or voice over speech, in the subject's immediate experience of his or her lifeworld.

In valorizing sound over meaning, which in turn elevates aesthetics over epistemology and pleasure over knowledge, Schreber is hardly alone among nineteenth-century aestheticians. Indeed, as I will suggest in this essay, with its talking birds, preoccupation with repetition, insistence on the sonic materiality of language, and the liberating potential of puns, *Memoirs of My Nervous Illness* has many affinities with the stories, poems, and critical writings of Edgar Allan Poe. After all, though Schreber recalls hundreds of birds, including woodpeckers, sparrows, and blackbirds, that overstimulated his nerves with "mechanical phrases learnt by rote," Poe's single croaking raven has become the most famous avian emblem in modern literature for imagining one suffering from the estrangement of sound and sense. When the mad, enervated speaker of "The Raven" begs the creature to stop repeating "Nevermore" and "take thy beak out of my heart" (86), is it not possible, placing this scene beside the bird-penetrated protagonist of the *Memoirs*, to see both Poe and Schreber as an absurd Promethean pair who, excluded from the symbolic order, are sentenced to endure monotonous torture from the vibrating needle of a bird's beak?

Schreber and Poe provide a phonosymbolic means for working through, and perhaps even tentatively evading, the automatization of linguistic experience. In Poe's "The Philosophy of Composition," the infamous 1846 essay that adopts machine-age rhetoric to explain how "The Raven" was "constructed" and brought to "completion with the precision and rigid consequence of a mathematical problem" (1375), Poe describes ravens, parrots, and other birds with the power of mimicry as "non-reasoning creature[s] capable of speech" (1378). Such a phrase also describes the droning automaton at the center of Poe's 1839 story "The Man That Was Used Up" and the various "miraculous" birds that assail Schreber's nerves. All will serve here as figures dramatizing the break between signifying utterance and signified meaning that is amplified via modern forms of mass communication. When Schreber blames the mechanization of his thinking on the birds and their "molesting him all day long with unconnected phrases," he also denounces their monotonous presence as an "*enormous infringement of man's most primitive rights*" (202). Simply put, both writers seek to reclaim some aspect of "man's most primitive rights" through the poetic/political redirection of verbal experience from *logos* into *phone* and, by extension, thought into feeling. To borrow from the case history with which I began, if Poe and Schreber experience words as dead letters emitted by Xerox machines passing for human beings, they also attempt to resuscitate cognitive and emotional deadness by simulating moments made stimulating by acts of sonic imitating.

Both writers, we will see, symbolically employ nonreasoning creatures capable of speech and engage with the power of punning in their writing to create dissonant cognitive spaces where the gap between sound and sense is laid bare. As the pun is the trope that effectively sets sound adrift by untethering it from unilateral meaning, punning presents a vehicle for depicting the free play of sound before it is captured, bound, and disciplined by the dominant speaking regime. Thus, this essay will ultimately conclude with an attempt to come to terms with the surprisingly disruptive power of punning, using Schreber and Poe's texts as models for the political confrontation of the emancipated voice and the disciplinary mechanisms of speech.

Schreber's Birds and the Political Order of Nonsense

In "The Philosophy of Composition," Poe claims that a writer should begin his work at the end. With this advice in mind, I'd like to follow the implications of voice as it appears in Schreber rather than Poe. My ultimate reason for doing so is because I believe that Schreber's homophony provides a new way of reading Poe's wordplay in "The Man That Was Used Up." If, as Borges famously noted, writers both create their own precursors and modify our conception of past texts, I'd like to consider how Schreber can modify the way we read Poe's writings when they are engaged as precursor texts to the *Memoirs*. In particular, Schreber's relationship with "the birds created by miracle" (190) helps us reimagine Poe's relation to the many nonhuman voices in his poetry and prose.

First, however, it is conceptually necessary to distinguish bare voice (*phone*) from meaningful speech (*logos*), since all that follows hinges on the difference. As Mladen Dolar helpfully defines it, "the voice (in its linguistic aspect) . . . is *what does not contribute to making sense*" (15). In this regard, it can be understood as a surplus of nonsense excluded from the symbolic order yet attendant, like the mute ghost of Banquo, at every speech act. Dolar and Jacques Rancière both note that in the *Politics* Aristotle excludes creatures given only voice from the political community of creatures given speech. "The supremely political destiny of man," Rancière writes, "is attested by a sign: the possession of the logos, that is of speech, which *expresses*, while the voice simply *indicates*" (2). Dolar adds that "mere voice is what animals and men have in common, it is the animal part in man," and so is effectively relegated the realm of "prehistory" by Aristotle (105). In other words, for both Dolar and Rancière, the exclusion of voice from the symbolic order of representation is the founding gesture of politics. I would like here to recall

Schreber's protest against the incessantly talking birds, which he claims infringe upon his "most primitive rights," especially the "natural right of mental relaxation, of temporary rest from mental activity through thinking nothing" (197). This "primitive" and "natural" right to think nothing is only achieved in the *Memoirs* when Schreber leaves the state of compulsive thinking and enters a state of mere voice. To reclaim his humanity, then, Schreber seeks a return to that "animal" and "prehistorical" state where speech has yet to triumph over voice. Similarly, Michael Ziser has shown that Poe's interest in the soundworld of creatures like ravens and orangutans reveals his attunement to the "zoosemiosphere," a "transspecific" space of undifferentiated voice and speech that affirms rather than denies continuity between animal and human life systems (11). Ultimately, then, Poe and Schreber's phonic turn away from "making sense" recodes animal-like nonsense, contra Aristotle, as belonging to political life.

Before revealing how and why he was periodically revitalized by accessing the surplus of voice traditionally excluded from the order of representation, Schreber first emphasizes the importance of escaping the "compulsive thinking," or the impulse of "having to think continually" (197), which the talking birds produce in him. Schreber believes in "man's natural right to be master of his own nerves" (55), but this is impossible because the vibrations of the birds' repetitive utterances unrelentingly assault his body and mind. They "reel off without knowing the sense of the words" whole "phrases learned by rote" that enter his nervous system as "long threads into [his] head" (192, 200). That is, the bare sounds emitted from the birds, what he calls the "poison of corpses" (190), are figured as a kind of vibrating electrophysical contaminant, not the signs of a language.

Furthermore, Schreber maintains that the birds repeat their phrases as a kind of affective prophylactic, since they immunize the poisoners from emotionally connecting with their victim, "as if they entered into me blindfolded or with their natural capacity for feeling somehow suspended" (192). Increasingly, the phrases that the various woodpeckers, blackbirds, and sparrows repeat take the form of questions and incomplete sentences that Schreber for his part is expected, respectively, either to answer or complete. The Althusserian analogy Schreber uses to illustrate his plight is that of an adult standing in a classroom who must listen to the questions a schoolmaster asks his pupils. The adult, by memory of his own childhood training, cannot resist silently answering the questions in his head (200). Thus denied mastery of his own nerves, Schreber is mastered by outside voices that are paradoxically experienced as speaking inside of him. Neither his thoughts nor his feelings are his own.

Just as Schreber's mind is in the process of being reprogrammed to think incessantly, he describes his body being transformed into that of a woman so that both God and he can sexually enjoy themselves incessantly. In his reading of the *Memoirs*, Freud developed his theory of paranoid psychosis by emphasizing how Schreber's fantasy of feminization and penetration by God betokens a case of homosexual desire that the paranoid subject wishes to suppress, substituting in its stead an elaborate persecution complex. This process of "feminization" involves becoming passive in body as well as mind, and so for Freud, the rote drilling of Schreber by the birds into thinking compulsively without regards to meaning amounts to the psychotic subject becoming like a "young girl." "In a carping mood," Freud writes, "people often compare young girls to geese, [and] ungallantly accuse them of having the 'brains of a bird,' declare that they can say nothing but phrases learnt by rote, and that they betray their lack of education by confusing foreign words that sound alike" (111). In a reading explicitly aimed against both Freud's misogyny and the reduction of religious experience to homosexual panic, Ida MacAlpine compares these birds to the figure of the dove as emblem of God the Father's creative word in Christian mythology. Jacques Lacan wittily suggests that rather than thinking of these birds as Holy Ghost–like hosts of the paternal logos, they are best imagined as the sort of dove a magician pulls out of his sleeve, the "meaning" of which is constituted solely by the sudden but meaningless arrival of its appearance (194).

In *My Own Private Germany*, however, Eric Santner recognizes Schreber as a representative subject of the disciplinary society and thus contends that the *Memoirs* provide its readers with a "secret history of modernity." For Santner, Schreber's vivid and bizarre narrative of abjection and feminization by an immanent yet absent power allegorizes the inscription of the modern subject into impersonal systems of bodily and mental control, interlocking apparatuses that Schreber calls "The Order of the World." Thus deviating from Freud's sexist hermeneutic while preserving much of its structure by substituting "feminized subject position" for "young girls," Santner instead posits the bare sonic emissions of these mechanically speaking birds as the hollow signals of a modernity characterized by its "despiritualization of language" (74). Indeed, one may look at the schoolroom atmosphere of their compulsive "interrogations" as figuring a hallucinatory drama of interpellation into an ideological state apparatus, the goal of which is to create the docile body and controlled mind of a successfully disciplined subject. Santner further contends that these avian tormenters might be fruitfully read as literary kindred to those seemingly powerful yet utterly disempowered angels of bureaucracy that inhabit Kafka's ultramediated world of "semihuman copyists, secretaries, and assorted servants" (75). Among many other things, one might call the compulsively driven Schreber a Bartlebyan captive of mechanically reproduced language, of ideologically programmatic language, an inmate in the prison house of uninspired, uninspiring language that George Orwell once compared in his own avian terms, and perhaps with his own sexist assumptions, to a "prefabricated henhouse" (357).

In a later study, Santner uses the Schreber case as a vehicle for understanding the loss of the modern individual's "capacity to cathect to—to inhabit, occupy, erotically invest in—the signifiers representing him in the world" (*Royal Remains*, xiv). As we have seen, this failure to cathect is allegorized in the *Memoirs* through the bird's use of meaningless words to make themselves immune to feeling anything while they unthinkingly torment Schreber. Yet a total lack of erotic investiture in the signifiers that one uses to represent oneself never fully arrives in the *Memoirs*. So here at the end of this section, I would like to focus on how Schreber renews affective attachment to the signifiers both he and his unthinking persecutors use to represent their presence to one another. He does so, I argue, at the phonic rather than semantic level. And when he does, he generates a fleeting community of captors and captive out of one shared affective response to a voice without meaning. Schreber describes the birds as consisting of the remnants of the souls of human beings. When phonically waylaid into experiencing what Dolar calls "the voice and nothing more," the soul fragments are briefly restored to full soulfulness. At such moments, "man's most primitive rights" can be enjoyed and even shared.

Schreber recalls two ways that he resisted the "phrases learned by rote" poisoning his body. One is reciting poetry. After writing that "playing the piano and reading books and newspapers" (203) have the power to put the voices momentarily at bay, Schreber suggests that one of the best ways to subvert the birds' incessant and meaningless compulsion to think is by "committing poems to memory":

> I learnt a great number of poems by heart particularly Schiller's ballads, long sections of Schiller's and Goethe's dramas, as well as arias from operas and humorous poems, amongst others from Max and Moritz, Struwelpter and Spekter's fables, which I then recite on the quiet verbatim. Their value as poetry naturally does not matter; however insignificant the rhymes, even obscene verses are worth their weight in gold as mental nourishment compared with the terrible nonsense my nerves are otherwise forced to listen to. (203)

Perhaps it is best, in the spirit of Santner's reading of the *Memoirs* as providing a secret history of modernity, to think of Schreber as a good modernist literary critic. As a kind of countermagic to the sort of rote, ideological language offered by the talking "birds created by miracle" (190), Schreber deploys, as many Euro-American literary critics of the twentieth century argued we should, the freed utterances of poetic language. Yet even this is not enough. Schreber bemoans the fact that the bird's voices reappear as "miracles aimed at scattering my thoughts [and that] act on my nerves and make it impossible to find the continuation of a poem learnt by heart" (203). Even the dulcet tones of Schiller and Goethe are overtaken by the torrent of relentless, repetitive interrogation performed by some hidden power's unthinking yet garrulous agents.

While the recitation of poetry learned by rote proves only a fleeting balm from the tyranny of phrases learned by rote, Schreber demonstrates another technique he developed to oppose his sonic torturers. He writes that

> if, while reeling off the automatic phrases, [the birds] perceive *either* in the vibrations of my own nerves (my thought) *or* in speech of people around me, words which *sound* the same or similar to their own phrases, they apparently experience surprise and in a way fall for similarity in sound; in other words, the surprise makes them forget the rest of their mechanical phrases and they suddenly pass over into *genuine* feeling. (192)

That is, homophony allows Schreber to gain a brief respite from mechanically reproduced speech as well as a means by which the birds, those attenuated scraps of humanity conscripted into God's service, are capable of experiencing actual emotion. They "have a natural sensitivity for *similarity of sounds*"; indeed, the sounds need not [even] be completely identical . . . as in any case the birds do not understand the *sense* of the words" (192). Therefore, they misrecognize, for example, "Santiago" for "Carthago" and "Chinesenthum" for "Jesum Christum" (192). What matters is that an aesthetic sensitivity to sound rather than a logical apprehension of sense momentarily sets the signifier in flight away from repetition and into the shared emergent event of "genuine feeling."

As I will suggest, in his story "The Man That Was Used Up," Poe uses puns, those figures of speech that sever individual sound from unilateral sense, in much the same way that Schreber does in his *Memoirs*. Also like Schreber, Poe will link mental and bodily captivity to the operations of mechanically repeating sound that masquerades as language. By foregrounding the gap between sign and meaning via a series of (often forced) puns, Poe locates a space before *phone* is subordinated to *logos*. There, to use Schreber's phrases, the "terrible nonsense" and "dreary monotony" (38) of compulsively (re)produced language in service to a symbolic regime's political interests can be short-circuited and perhaps even rewired. First, however, I would like to consider some of Poe's own musings on his use of a talking bird—one who also serves a kind of torture function—and how it figures into his understanding of the sonorous confrontation between language and voice.

Poe's Bird and the Poetic Order of Nonsense

What do Schreber's birds and Poe's raven have in common? To echo the title of Dolar's book: *A Voice and Nothing More*. That is, while both kinds of bird can emit sounds that formally approximate linguistic signifiers, such sounds cannot be said to mean anything properly, since such birds as ravens and parrots are, as Poe writes in "The Philosophy of Composition," "non-reasoning creatures capable of speech" (1378). Early in that essay, even before his selection of the raven as his primary symbol, Poe claims that he chose to include the word "Nevermore" in his poem's refrain because of its combination of the long *o* and *r* sounds, respectively "the most sonorous vowel" and "the most producible consonant" available to the poetic craftsman (1378). That is, he's more interested in "Nevermore" and the poem in general for its aural rather than discursive, phonic rather than logical, properties.

Nevertheless, Poe does admit that his desire to have the same one word repeat time and again in the poem proves rationally vexing, but only because he allowed himself to stray into the "pre-assumption that the word was to be continuously or monotonously spoken by a human being" (1378). Poe solves this logical problem of representation ("Who but a madman would repeat the same word over and over again?") by conjuring forth a nonhuman vocalist, first "a parrot" and then "a Raven," ultimately deciding upon the latter since it is "equally capable of speech, and infinitely more in keeping with the intended *tone*" (1378). The difficulties of *logos* in poetry, that is, are solved by an authorial turn to *phone*. After all, as an aesthetic theorist, Poe defends art's right to exist outside of moral or logical concerns, claiming in "The Poetic Principle" that the literary artist can never "reconcile the obstinate oils and waters of Poetry and Truth" (1436). So then, for one who desired to write poetry that could achieve the beautiful effects of pure sound, what better device for delivering sonorous vowels and producible consonants than a creature biologically unequipped to understand the meaning of "what its voice makes? This does not, however, seem to allay the anxieties of the speaker of "The Raven," who by poem's end is begging the monotonous bird to "take thy beak out of my heart" (86).

Yet if Schreber and the poetic speaker of "The Raven" are both victimized by repetitious voices described as violating the body's boundaries, one notes at least one major difference. Poe's speaker misrecognizes the bird's nonsense as somehow concealing secret depths of meaning. The speaker mourning Lenore by hearing the bare sounds of her name in "Nevermore," it seems, is Poe's ideal reader, one aurally susceptible and rendered prostrate by the dictatorial triumph of meaningless sound. Poe's perfect poem, like Schreber's "Order of the World," is an artifact indicative of expertly administered nonsense. As Poe makes clear in "The Philosophy of Composition," though, if the speaker of "The Raven" resembles a good reader of poetry, he does not resemble a good poet, who, it is claimed, despite popular belief, does not compose verse in a state of frenzy but rather in the measured mood of a gentleman machinist laboring methodically on some ingenious and wholly unnecessary gadget. Meanwhile, unlike the speaker's inability to see that the bird knows not what it says, the paranoid, psychotic Schreber recognizes the complete nonsense of the repetitive, wordlike sounds brought forth from his persecutors. As I suggested earlier, if we accept Santner's claim that Schreber's illness generates fantasies of persecution at least partially from his experience of language as a mass-produced, desacralized entity in service to the sociopolitical machinery of discipline, we might also see his use of poetry as a weapon to shoo away the chattering birds as a standard aesthetic strategy of Euro-American modernism that opposes the carefully crafted language of original expression (often coded as masculine) against the commodified language of mass communication (often coded as feminine).

For example, before scholars like Terrence Whalen, J. Gerald Kennedy, and Jerome McGann helped debunk it by highlighting his shrewd negotiation of the popular marketplace, the classic "Poe myth" dictated that Poe's was an original genius tragically wasted by a life of hackwork in the periodical press. Poe himself helped build this myth, stating in the 1845 Preface to *"The Raven" and Other Poems* that material conditions made it impossible to pursue a fulltime poetic vocation: "Events not to be controlled have prevented me from making, at any time, any serious effort in what, under happier circumstances, would have been the field of my choice" (18). His best poems, seen in this light, are thus even more worthy of our admiration because they stand heroically above the formulaic popular genres he was supposedly sentenced by fate to traffic in given the vulgarity of American tastes. Poe desired to write poetry that is nonutilitarian, beautiful, original, and subversive, while the language of magazines is a blunt communicative tool: homely, secondhand, a substance designed to cement the sociopolitical order of the bourgeois status quo. Yet before we are too quick to valorize the originality of poetry over the clichés of popular discourse, Eliza Richards has shown that poems like "The Raven" contributed to a nineteenth-century aesthetic, most often linked to woman's poetry, which celebrated mimicry as more important to poetic success than originality. After all, Poe, as his monotone raven proves, was hardly one to belittle the pleasures of repetition. One need think only of the hypnotic metrics of poems like "Ulalume" or "Annabel Lee," the plagiarized passages of *Arthur Pym*, or the pendulums, clocks, automata, vortices, *Blackwood* articles, beating hearts, bells, apes, and other incessantly repeating forces, animals, machines, texts, and organs that populate his work.

Like Schreber, Poe's resistance to the mass mechanization of language stages itself as a symbolic drama featuring disruptive figures of sonic repetition like ravens and parrots, those nonreasoning creatures capable of speech who sonically lay bare the distinctly phonic dimension of verbal communication. Such symbolic figures momentarily restore, to use Schreber's phrase, a sense of "genuine feeling," an ephemeral cognitive effect that Poe associates with poetry's aesthetic triumph over logic. Yet mechanical reproduction also seems such a constituent feature of how Poe understands life and language that he sometimes appears not wholly confident in the ability of art to distinguish meaningfully among rational animals and nonreasoning creatures capable of speech. As I will argue, in the short story "The Man That Was Used Up" the very sort of repetition compulsion that tends both to structure and provide the subject for much of Poe's writing can also work to expose a surprisingly uncanny relationship between ideological and poetic discourse.

Poetry and/or Ideology: The Pleasure of Merely Repeating

In "The Philosophy of Composition," Poe recounts how he wanted a "keynote" in his poem that would echo throughout, so he decides on the poetic device of a refrain, the "pleasure of which is deduced solely from the sense of identity—of repetition" (1377). Having committed to the use of a refrain, he then chooses for it a word that contains the phonetic combination of a long *o* and *r*, "Nevermore," which also evokes the mood of melancholy he wishes to summon in the poem. However, Poe admits he was initially perplexed at how he might justify, within a narrative poem, the repeating of one word over and over again:

> In observing the difficulty which I had at once found in inventing a sufficiently plausible reason for its continuous repetition, I did not fail to perceive that this difficulty arose solely from the preassumption that the word was to be so continuously or monotonously spoken by a human being—I did not fail to perceive, in short, that the difficulty lay in the reconciliation of this monotony with the exercise of reason on the part of the creature repeating the word. Here, then, immediately arose the idea of a non-reasoning creature capable of speech, and very naturally, a parrot, in the first instance, suggested itself, but was superseded forthwith by a Raven as equally capable of speech, and infinitely more in keeping with the intended tone. (1378)

In this passage, while Poe points out the seeming absurdity of a reasoning creature who would repeat the same sound over and over again, he also makes a case for the nonlogical character of poetic language itself. Thus, just as Poe takes on the persona of a hyperrational explicator of how a poem works, he also suggests that the "key-note" of his poem owes nothing to reason at all. To repeat, it is a voice and nothing more.

Yet, if in composing "The Raven" Poe finds the notion of a human being repeating the same phrase over and over again ridiculous, in "The Man That Was Used Up," the sight and sound of human beings who repeat the same words over and over again is the very focus of its concerns. For while there are no talking birds in this story, there is the monotonously speaking social celebrity Brevet Brigadier General A. B. C. Smith, who we learn is a cyborg constructed out of designer prosthetics, having lost nearly every single body part in the genocidal wars of antebellum Indian removal. The only original piece of human left on Smith is a bit of tongue. And with this, aided by an oral prosthesis, he infects the subjects who adore him with progress-mongering clichés that spread like a sonic virus, so that in the end, everybody repeats the handful of sounds they hear in A. B. C.'s voice. To invoke Schreber yet once more, the sounds coming out of Smith's mouth are best described as the poison of corpses. For while General Smith's body and language may indeed be dead, the unkillable surplus of his voice remains intact and quite capable of occupying its listeners. Thus, in contrast to Schreber, for whom the nonsensical surplus of voice creates a free space for affective rapprochement with the world, in Poe the nonsense of voice is amplified across an imperial system of violent nonmeaning where reason (along with the Indians "removed" on its behalf) has been driven from the Western polis entirely.

Borrowing the film theorist Michael Chion's notion of the "acousmatic voice," Mladen Dolar describes how voice-capturing technologies like the phonograph detach sonic utterance from its origin in mass culture. Helping illuminate the term, Dolar writes: "The acousmatic voice is simply a voice whose source one cannot see, a voice whose origin cannot be identified, a voice one cannot place. It is a voice in search of an origin, in search of a body, but even when it finds a body, it turns out that this doesn't quite work, the voice doesn't stick to the body, it is an excrescence which doesn't match the body" (61). "The Man That Was Used Up" can be productively read as a parable of the acousmatic voice, which Dolar maintains haunts modernity, a disembodied form of commanding utterance that masters its auditors even as the originating site of that mastery is left in mystery. In order to provide support for such an allegorical reading, however, it is worth looking at the tale itself.

What bare plot there is in "The Man That Was Used Up" involves the quest of an anonymous, nervous man suffering from amnesia who wishes to meet General A. B. C. Smith, a figure who has sent the entire social world of the tale into talking. Smith was taken prisoner and tortured by Indians after what one character calls "a tremendous swamp fight" (309). Such a fight probably refers to the Second Seminole War, which was being waged during Poe's composition of the tale by the U.S. government against not only the Native American inhabitants of Florida but also their African-descended allies, the Black Seminoles, many of whom were descendants of slaves who had escaped from their English and Spanish masters as early as the seventeenth century. Yet Poe's interest in historical fidelity seems indifferent at best. The subtitle of the story is "A Tale of the Late Bugaboo and Kickapoo" campaign. And though the indigenous Kickapoo fought beside Tecumseh against white encroachment into the Mississippi River Valley, the pseudoterm in the subtitle, "Bugaboo," is perhaps a better indicator of Poe's notion of Indians, as bugaboo is a vaguely onomatopoeic nonsense word that disavows the real threat of nonwhite opposition against the growth of American power, reducing the reality of Seminole, Kickapoo, and maroon resistance into the sort of virtual boogeymen that white parents might invoke to scare their children into good behavior.

In the narrator's first meeting with Smith, which amounts to little more than the latter's monologic assault on him (a repetitive experience the narrator admits to finding pleasurable), the voice of global white empire says:

> There is nothing at all like it . . . we are a wonderful people, and live in a wonderful age. Parachutes and rail-roads—mantraps and spring-guns! Our steam-boats are upon every sea, and the Nassau balloon packet is about to run regular trips (fare either way only twenty pounds sterling) between London and Timbuctoo . . . There really is no end to the march of invention. (310)

After this meeting, the narrator seeks new intelligence about Smith. A Miss Tabitha T. says to him, "'Smith!—why, not General John A. B. C.? Bless me, I thought you knew all about him! This is a wonderfully inventive age! Horrid affair that!—a bloody set of wretches, those Kickapoos!—fought like a hero—prodigies of valor—immortal renown'" (311). Still seeking information, he asks Miranda Cognoscenti for some: "'Smith!' said she, 'why, not General John A. B. C.? Horrid affair that, wasn't it?—great wretches, those Bugaboos—savage and so on—but we live in a wonderfully inventive age!—Smith!—O yes! great man!—perfect desperado—immortal renown—prodigies of valor!'" (312). And so forth. Each person hurls the same set of clichés disseminated by Smith and about Smith. These subjects, hosts to parasitic dead letters no less than Schreber's birds, are incapable of generating their own felt expressions or producing fresh meanings for the sounds that come out of their mouths.

Yet upon hearing each person speak about Smith, the narrator gets more impatient to see him again. He makes an appointment to see the man at his home:

> It was early when I called, and the General was dressing, but I pleaded urgent business, and was shown at once into his bedroom by an old negro valet, who remained in attendance during my visit. As I entered the chamber, I looked about, of course, for the occupant, but did not immediately perceive him. There was a large and exceedingly odd looking bundle of something which lay close by my feet on the floor, and, as I was not in the best humor in the world, I gave it a kick out of the way.
>
> "Hem! ahem! rather civil that, I should say!" said the bundle, in one of the smallest, and altogether the funniest little voices, between a squeak and a whistle, that I ever heard in all the days of my existence. (314–315)

The acousmatic voice coming from out of the bag proves to be that of Smith, whose body is scattered into pieces. The narrator watches, while piece by piece, the general is constructed by a black valet, Pompey, whom the voice from the bag addresses in the race-baiting, belligerent tones of human commodification:

> Now, you dog, slip on my shoulders and bosom. Pettit makes the best shoulders, but for a bosom you will have to go to Ducrow . . . Now, you nigger, my teeth! For a good set of these you had better go to Parmly's at once; high prices, but excellent work. I swallowed some very capital articles, though, when the big Bugaboo rammed me down with the butt end of his rifle. (315)

The last piece of the general's assembly is his palate. Once this is in place, the narrator notes of Smith that "his voice had resumed all that rich melody and strength" (316) he heard upon their first meeting and that had so impressed itself upon his many social admirers. "'Damn the vagabonds!' said he, in so clear a tone that I positively started at the change, 'Damn the vagabonds! they not only knocked in the roof of my mouth, but took the trouble to cut off at least seven-eighths of my tongue'" (316).

Throughout the tale, Smith, despite the masculine feats of Indian-killing prowess ascribed to him, is cast in a feminized subject position. Upon the narrator's first glimpse of Smith, his erotic gaze results in an elaborate blazon that extols the perfection of the general's every body part, from his Brutus-like head and broad shoulders to his femur and calf. Yet, by the end of the story, he learns that each of those parts he has minutely praised is a fake. The robotic general's reassembly, represented as a woman's getting dressed for public in her boudoir, effectively deconstructs the previous paean to his impressive physiological makeup. D. H. Blake has written that "The Man That Was Used Up" is Poe's reflection on the Indian captivity narrative genre, which in his hands demonstrates how the white male body is physically violated and then textually reconstituted as a feminized body whose ability to tell the "truth" of his story is set aside to instead serve the ready-made ideological purposes that the Indian captivity narrative traditionally has served. Just as he undergoes a change into a feminized subject position—remember Schreber too is undergoing a slow transformation into a woman—the truth of the warrior's lived experience is transmitted as ideology in the form of automatic phrases, and the general's banal boilerplate performances make him a celebrated participant in the female gossip circles and drawing-room parties that constitute social life in the tale. It is important to recognize that the general's voice is not his own. The "damn vagabond" aspect of it is that it is ultimately a *vox populi* impossible to fix in any one place. Indeed, all the men and women repeat like parrots (or ravens) the general's racist and banal celebratory pronunciations on the benefits of Western progress and mechanical invention. The origin of so much Manifest Destiny and warmongering is distributed across a discursive field–made–echo chamber, located at everywhere and nowhere at once.

Poe's implication in "The Man That Was Used Up" is that American public speech is ontologically used up, a generalized ABC of expression resulting in what Schreber called "compulsive thinking," reproduced among the voices of those who speak it through the sheer repetitive force of its irresistible refrain. Thus in its discarding of reason, made concrete in the automaton General A. B. C. Smith and his dissemination of mechanical phrases, in its severing of linguistic sign from material reference, Poe's depiction of ideological discourse, I am arguing, uncannily resembles Poe's depiction of poetry. Both result in a kind of entranced subject mastered by the tones of nonsense. Poe gestures toward this potential paradox in his tale by surrounding his narrator with instances of poetic utterance in direct sonic conflict with the language of Smith's admirers. At these moments, I will suggest, the appearance of puns is crucial. In "The Man That Was Used Up" they demonstrate how the surplus materiality of the voice without meaning, rather than reinforcing and amplifying the voice of the status quo, can also become the raw sonic stuff out of which one might begin to take up an alternate language game than the one that the Western course of empire has traditionally played.

In the middle of each conversation that the story's narrator has with one of the general's brainwashed fans, verbal exchange is halted by a pun on the word "man" just as the narrator is about to hear that Smith is the titular "man that was used up." For example, during a performance of *Othello*, a woman begins to say that the general is "the man that was used up": "'Bless my soul!' she says, why, he's the man—'" (312). Suddenly, however, the actor playing Iago raises his voice to drown hers out:

> "—*man*dragora
> Nor all the drowsy syrups of the world
> Shall ever medicine thee to that sweet sleep
> Which thou ow'dst yesterday."
>
> (312, emphasis mine)

This piece of speech from Act 3 is given just after Iago compares the lies he has told Othello concerning Desdemona's faithfulness to his administration of poisons that no drug can cure. Iago, like Poe's cyborg, is not what he seems, and his words, like Schreber's "poisons" or Smith's clichés, seem to work their pernicious effects as much on the body as the mind. There's also an element of bad interracial faith going on, as the white Iago's words are straightaway destructive to the black Othello. In another instance of punning in the story, "man" is misheard by a woman at a party as "Captain Mann" (315), a person who helped finance a working-class savings bank that failed when it printed money in excess of its capital. It is another scene of bad faith, this time between conning bankers and credulous workers. So these puns on that tricky word "man," a pun anyway, since the sound "man" can name a particular individual or a general entity (think of Schreber's insistence on "man's primitive rights"), illuminate the asymmetrical political field in which some dominant race or class verbally schemes to exploit one perceived as other, and hence subordinate, to it. "Man," I am suggesting, is a sound that emerges from this political asymmetry.

Drawing on Freud's treatment of puns in *Jokes and Their Relation to the Unconscious*, Dolar writes:

> Words, quite contingently, sound alike, to a greater or lesser degree, which makes them liable to contamination; their mutual sound contacts can transform them, distort them, be it by retention, the inertia of certain sounds, their momentum by which they influence what follows, or by anticipation of certain sounds which influence what precede them, or by various modes of substitution. In this contamination a new formation is born—a slip which may sound like nonsense but produces the emergence of a newer sense. (140)

Schreber described this "emergence of a newer sense" as the "genuine feeling" that suddenly coursed through the bodies of the mechanically speaking birds that tormented him when he produced homonyms to confuse their chattering. Short-circuiting the dispersal of automatic language by affectless mediators, puns for both Dolar and Schreber lay bare a prepolitical political space where the inscription of an individual into a symbolic order is temporarily brought to the moment of a standstill. Similarly, Poe brings his readers to such moments in the series of puns that happen throughout "The Man That Was Used Up," especially the one that I'd like to explore next. As in the case of "mandragora" and "Captain Mann," Poe's puns on "Man" open up a space where the complicit mechanics of domination and subordination are laid bare at the site of a single utterance.

The Enclosure of Speech and Its O-punning

Let us return once more to the scene of Poe's narrator attempting to learn about General Smith. "'I shall have to sit down and enlighten you,'" says a character named Mrs. Pirouette. "'Smith! Why, he's the man—'" at which point another woman named Miss Bas-Bleu overhearing the conversation shouts, "'Man-Fred, I tell you!'" She goes on:

> "Did you ever anybody hear the like? It's Man-Fred, I say, and not at all by any means Man-Friday." Here Miss Bas-Blue beckoned to me in a very peremptory manner; and I was obliged, will I nill I, to leave Mrs. P. for the purpose of deciding a dispute touching the title of a certain poetical drama of Lord Byron's. Although I pronounced, with great promptness, that the true title was Man-Friday, and not by any means Man-Fred. (313)

This series of misunderstandings hinges on the pun constructed out of "man," which in this context can refer to Byron's brooding and incestuous magus, Crusoe's Christianized cannibal, or the automaton Smith. Mrs. Pirouette promises enlightenment, but confusion reigns, as the name of the paradigmatic figure of "I will not serve" in English literature (Manfred) is mistaken for the name of the paradigmatic figure of "I will serve" in English literature (Man-Friday). Suddenly, that is, the Übermensch and the colonized Indian are mistaken for the same man. Following Dolar's characterization of a pun as adulterated sonic material capable of creating new epistemological formations out of its very contamination, I would like to construct a reading of this pun more closely and suggest that Poe's contamination of the sound /man/ opens up a space for revisiting the word "man."

In "The Philosophy of Composition," Poe critically separates verse, which should be read in one setting to achieve unity of effect, from prose narrative, which can be read in multiple sittings, since its effects are fragmented and various. His example of prose narrative that he contrasts with "The Raven" is Defoe's *Robinson Crusoe* (1375). Thus one might take the figures of Manfred and Crusoe's Man-Friday as metonyms for, respectively, poetry and prose. Poe also recounts how he discards the idea of a parrot as his talking bird for that of a raven, since the former's bright motley would be out of touch with the poem's desire to create a unified sense of melancholy. One literary parrot that Poe certainly knew was Crusoe's Poll, the bird that Defoe's enterprising colonist teaches English words to before doing the same to Friday. Laura Brown has claimed that stories regarding talking birds, particularly parrots, are foundational fables of modernity, particularly in a New World context, where the speech of a parrot like Poll mimics how the colonized subject experiences taking up the colonizer's language, as Crusoe's Friday does, without, or so the Eurocentric fable goes, comprehending its meaning (245–254). Voice, Rancière reminds us, only indicates animal existence; language alone expresses humanity. Politically speaking, the humanity of the white male subject of modernity excludes the voice of the other in the many derivative forms it can take: the female voice, the enslaved voice, the colonized voice, the animal voice, the mad voice. In this vein, in the *Memoirs* Schreber suspects his birds are emissaries of "the Eternal Jew" (61).

We have already seen in "The Man That Was Used Up" that Iago's poisoning of the African Othello's mind mimics the manner in which Smith spreads the virus of his racist, imperial, and technophiliac utterances. Yet, if we consider the general's speeches in the manner of the acousmatic voice, we recognize that his sonic repetitions belong to no one in particular, even as they form a general consensus of belief regarding how the world should be politically managed. Though Poe's misogyny runs rampant in the tale (most of the narrator's mechanically speaking interlocutors are female), the general's speech infects the voices of men as well. And as I've argued, the general himself is placed in a feminized position throughout. Even the master/slave relationship is inverted in the general's boudoir, as Smith's racial invective cannot hide that his power depends on the labor of his black servant Pompey. When, for example, Pompey checks if the fake teeth of his "master" are placed in his mouth properly, he is described as showing "the knowing air of a horse-jockey" (316). The slave rides the ridden master. These reversals in subject position are typical of Poe's text. Subjects all to the totalizing political logic of Western empire—Smith is "killed," after all, in a state-sponsored, racist war for territory—no one is safe from the historical effects that the nonsensical refrains of its ideology produces. "Man," that is to say, is always already an abject other, already a Man-Friday. Thus by bringing together Manfred and Man-Friday as one dual image of the concept of "Man," I am suggesting that Poe recognizes the dialectical relationship upon which master and slave depend. Furthermore, by locating this dependence at the voiced site of a pun that carries a surplus of extralawful meanings, Poe momentarily lays bare a site where the mechanisms of political exclusion and inclusion are thrown into relief.

All of which is to say, I am making a strong case for the importance of puns. Puns are tropes that operate phonically, and thus traditionally they have been subordinated by literary critics to those like metaphor and metonymy, which work at an epistemological register. Such critical arrogations of *logos* over *phone* effectively reproduce the founding political gesture recited by Aristotle, in which speech rather than voice is allowed to dominate. Yet this, I am suggesting, is why puns matter to writers like Poe and

Schreber and to critics like Freud and Dolar. In defining what he calls "the politics of the voice," Dolar compares the voice before it is excluded to serve the symbolic needs of a language to Giorgio Agamben's notion of a state of exception, in which sovereign power "can suspend the validity of the law and inaugurate the state of emergency" (120). Occupying a figural space of phonic and logical surplus, the nonsense of homophony exceeds the symbolic order it precedes and ultimately produces. Puns, therefore, according to Freud and Dolar and the argument of this essay, in the manner by which they isolate vocalization from unilateral semantic intention, mark a moment of emergence in which the structural enclosures of language are threatened by the surplus they fail to contain.

Thus, in conclusion, I would like to fashion my own punning performance, constructed from a reading of the texts that Poe himself brings together in his pun on "Manfred" and "Man-Friday." I'll begin with Poe's contention that

> Having made up my mind to a *refrain*, the division of the poem into stanzas was, of course, a corollary: the *refrain* forming the close to each stanza. That such a close, to have force, must be sonorous and susceptible of protracted emphasis, admitted no doubt: and these considerations inevitably led me to the long *o* as the most sonorous vowel . . . (1378)

Poe imagines a sonorous space where the aesthetic pleasure caused by a sound overwhelms the ravenous desire to comprehend it. He supposes an auditor with some primitive need for a sensual experience through which, to echo Schreber, "genuine feeling" occurs. In this regard, Poe's "long *o*" recalls a passage from one of Manfred's speeches, in which the all-powerful, powerless hero says

> *Oh*, that I were
> The viewless spirit of a lovely sound,
> A living voice, a breathing harmony,
> A bodiless enjoyment—born and dying
> With the blessed tone which made me!
>
> (473, emphasis mine)

Byron's protagonist begins with the emotionally meaningful/logically meaningless interjection "Oh," producing what Poe called the most sonorous vowel sound possible. Manfred, like Poe, espouses the desire to inhabit a "living voice" spatially and temporally consubstantial only with the "bodiless" and "blessed tone" of its utterance. Both Byron and Poe, that is, seek enjoyment outside what can be reproduced: a "lovely sound" that perishes at the moment of speaking.

Meanwhile, Friday in Defoe's *Crusoe* reveals his own relationship to the sense and nonsense of "O." When describing his Man-Friday's religious condition, Crusoe explains that while the native had a notion of some supreme creator,

> He could describe nothing of this great person, but that he was very old, much older, he said, than the sea or the land, than the moon or the stars, I asked him then, if this old person had made all things, why did not all things worship him? He looked very grave, and with a perfect look of innocence said, "*All things do say O to him.*" (170–171)

In response, Crusoe takes it upon himself to teach Friday the contents of the Bible, to indoctrinate him into the tenets of the white man's religion, and in sum "to instruct him in the knowledge of the true God" (171). The savage's ejaculatory prayer of "O" must be rerouted from an oral pagan culture into the truth of Christianity's logos, made flesh, of course, in the Protestant colonialist's Bible. The /ō/ of the other's voice, as well as the putatively irrational, prehistorical condition it represents, that is, must be excluded from the political life of Crusoe's island.

One may go further and see how Crusoe's evangelization of Friday's pagan soul coincides with the dietary disciplining of his cannibal stomach. Friday is taught to prefer the flesh of goats to the flesh of other human beings. Crusoe's "man," then, figures a form of surplus enjoyment that transgresses one of the two great taboos of Western modernity: cannibalism. The other, of course, is incest, best figured by Manfred. As noted, Schreber calls one of the gods that torments him Ariman. In a footnote to his *Memoirs*, Schreber attributes this name to its appearance "in Lord Byron's Manfred in connection with a soul murder" (31). In his analysis, Freud associates this allusion to *Manfred* with Schreber's incestuous feelings for his father, displaced onto his psychiatrist Emil Flechsig, and then finally God/Ariman (120). Santner translates this insight into sociopolitical terms:

> Freud's crucial point is that soul murder is connected with incest; Flechsig-as-soul-murderer becomes a figure of *incestuous enjoyment*. The surplus power/influence that Schreber sees as emanating first from Flechsig and then from God is thereby linked to that most powerful and primordial of transgressive stains on the "lawful" structure of kinship relations. (*Germany*, 39)

Both Friday's cannibal "O" and Manfred's incestuous "Oh" give voice to surplus pleasure that only registers as nonsense within the heavily disciplined enclosure that *Memoirs of My Nervous Illness* calls "The Order of the World."

Ultimately, the phonic turn away from *logos* that I am tracing, to continue borrowing from Schreber's immensely useful terms of analysis, excessively asserts "man's most primitive rights." If, as I have argued, the raven and the automaton of Poe and the miraculous talking birds of Schreber—collectively grouped as nonreasoning creatures capable of speech—allegorize the disciplinary function of mechanized verbal repetition in the formation of the modern subject, both authors also deploy the sonorous excesses created through homophony as a means of rethinking the boundary between sense and nonsense, reason and feeling, politics and poetics. Finally, through bringing together a collection of literary texts under the sign of the phoneme /ō/, I have made a performative attempt to seize upon the surplus nonsense of Poe's pun on Manfred and Man-Friday and extend its poetic/political potential. Surely, it still seems lacking. The two identities they bring to light, the noble savage and the Faustian seer, are among the most well-worn refrains for symbolizing a desire for human liberation from automatization in the age of mechanical reproduction. Generally, however, they no longer signify for contemporary, all-too-posthuman readers except as the "poison of corpses," the racist and antidemocratic figures of past myth.

Schreber wrote that sometimes the talking birds would simply torture him with random interjections, so that he would feel, every so often, "an absolutely senseless 'Oh' thrown into my nerves" (198). I'd like the absolute senselessness of Schreber's "Oh" to perform the final opening gesture of this /ō/ punning, hoping then to give some sense of the voice haunting that other nonreasoning creature capable of speech called man.

WORKS CITED

Blake, David Haven. "'The Man That Was Used Up': Edgar Allan Poe and the Ends of Captivity." *Nineteenth-Century Literature* 57, no. 3 (2002): 323–349.

Brown, Laura. *Fables of Modernity: Literature and Culture in the English Eighteenth Century*. Ithaca, N.Y.: Cornell University Press, 2001.

Byron, Lord George Gordon. *Selected Poems*. Ed. Susan J. Wolfson. New York: Penguin, 2006.

Defoe, Daniel. *Robinson Crusoe*. Ed. John Richetti. New York: Penguin, 2003.

Dolar, Mladen. *A Voice and Nothing More*. Cambridge, Mass.: MIT Press, 2006.

Freud, Sigmund. *Three Case Histories: The "Wolf Man," the "Rat Man," and the Psychotic Doctor Schreber.* Ed. Phillip Rieff. New York: Touchstone, 1996.

Kennedy, J. Gerald, and Jerome J. McGann. *Poe and the Remapping of Antebellum Print Culture.* Baton Rouge: LSU Press, 2012.

Lacan, Jacques. *Écrits: A Selection.* Trans. Bruce Fink. New York: Norton, 2004.

Orwell, George. *The Orwell Reader: Fiction, Essays, and Reportage.* Ed. Richard H. Rovere. San Diego: Harcourt Brace & Company, 1984.

Poe, Edgar Allan. *Poetry, Tales, and Selected Essays.* Ed. Patrick F. Quinn and G. R. Thompson. New York: Library of America, 1996.

Rancière, Jacques. *Disagreement: Politics and Philosophy.* Trans. Julie Rose. Minneapolis: University of Minnesota Press, 1999.

Richards, Eliza. "Lyric Telegraphy: Women Poets, Spiritualist Poetics, and the 'Phantom Voice' of Poe." *Yale Journal of Criticism* 12, no. 2 (1999): 269–294.

Santner, Eric L. *My Own Private Germany: Daniel Paul Schreber's Secret History of Modernity.* Princeton, N.J.: Princeton University Press, 1996.

———. *The Royal Remains: The People's Two Bodies and the Endgames of Sovereignty.* Chicago: University of Chicago Press, 2011.

Sass, Louis A. *The Paradoxes of Delusion: Wittgenstein, Schreber, and the Schizophrenic Mind.* Ithaca, N.Y.: Cornell University Press, 1994.

Schreber, Daniel Paul. *Memoirs of My Nervous Illness.* Trans. and ed. Ida Macalpine and Richard A. Hunter. New York: NYRB, 2000.

Whalen, Terrence. *Edgar Allan Poe and the Masses: The Political Economy of Literature in Antebellum America.* Princeton, N.J.: Princeton University Press, 1999.

Ziser, Michael. "Animal Mirrors: Poe, Lacan, von Uexküll, and Audubon in the Zoosemiosphere." *Angelaki: Journal of the Theoretical Humanities* 12, no. 3 (2007): 11–33.

CHAPTER 6

Splitting the Lyric Lark; or, Dickinson's Music Box

Isabel A. Moore

a
technology
of

numbered
human
kind

scar
let
experiment

the
thing
itself

and
its
song

gush
after
gush

dickinson's
music
box

splits
the
lyric

lark
in
two

Hope is kind of like birds. In that I don't have any.

—PAUL LEGAULT, *The Emily Dickinson Reader: An English-to-English Translation*

Paul Legault's *The Emily Dickinson Reader: An English-to-English Translation* (2012) offers tongue-in-cheek renditions of the 1,789 texts numbered by R. W. Franklin in his reading edition of *The Poems of Emily Dickinson*. My epigraph is Legault's translation of Franklin's #314, which opens: "'Hope' is the thing with feathers -/That perches in the soul -/And sings the tune without the words -/And never stops - at all -" (ll. 1–4, 140). Legault characterizes his translation as an attempt to "rewrite [Dickinson's] poems (with their foreign beauty intact) in 'Standard English'" (8); we might characterize it as a loving instance of the poetics of tweeting more than parroting. Legault's #314 is a mere fifty-three characters to Franklin's 362, and its condensed and casual tone at once suspends—and in some cases offends—critical beliefs and generic sensibilities, both around Dickinson and around the genre in which we think she wrote.[1] In Legault's description, Dickinson has become "the father [*sic*] of American poetry" (7),[2] which poetry is inevitably believed to be lyric: mistaken as the first and sometimes eternal genre, rendered in splendid isolation, and inoculated from any context save the act of reading itself (Jackson, *Dickinson's Misery*, 6–8). Legault at once parrots and participates in the resurrection of Dickinson and her lyricized legacy, as in the mildly ominous but comic declaration with which he closes his introduction: "*Emily Dickinson used to exist. Now she's doing it again*" (8). That Dickinson continues to exist, zombie- or Elvis-like, in her work and apart from her historical context serves to perpetuate the assumption that Legault cites from Franklin's *Poems*: "This edition is based on the assumption that a literary work is separable from its artifact, as Dickinson herself demonstrated . . . there can be many manifestations of a literary work" (in Legault, *Reader*, 8). Certainly there can be as many manifestations as there are translations (or editions), but the deeper assumption in both Franklin and Legault is that what Dickinson wrote was *already* literature, and specifically lyric, rather than another kind of writing (as Virginia Jackson has shown). And below even this assumption lies the belief that something essentially lyrical is not only identifiable but can survive across historical periods and beyond its material manifestation. It thereby ensures that both Dickinson and the genre live on in spirit, whatever their material form. I characterize this critical faith in generic life as a kind of lyric vitalism: one that posits a vital lyric force before and beyond its manifestation or expression in a particular form or object.

This essay seeks to dissect critical faith in lyric vitalism through the occasion of Dickinson's "Split the Lark." Both Dickinson and lark are already subject to the critical belief or fallacy that Virginia Jackson identifies as hypostatization: the process whereby abstract concepts are given (misplaced) substance and treated as if they were material or concrete realities. As Jackson puts it in *Dickinson's Misery: A Theory of Lyric Reading*, just "as the hypostatized lyric had come to stand for poetry in general, so Dickinson had come to stand for the hypostatized lyric in general" (41–42)—a critical assumption "implicit in Dickinson's reception since 1890" (41). While I won't explicitly retrace the intricate critical and material history that Jackson has so ably made visible, I will explore the implicit vivification that accompanies hypostatization in recent readings of "Split the Lark." By vivification, I mean the way in which the abstract concept of the lyric is made real or material by critics largely in living objects, whether humans (Dickinson) or birds (lark). Just as poetry—already assumed to be all lyric, all the time—has been made concrete in the figure and fascicles of Dickinson through hypostatization, it has at the same time been brought to life in the figure of Dickinson's lark through a parallel process of vivification.

Against the vivification of hypostatized concepts and the critical faith that perpetuates it, I will attempt a form of skepticism in keeping with the matter—not to say the spirit—of "Split the Lark." To parrot Legault's phony "hope" and to counter any reading of Dickinson and lyric as birds of a feather, neither Dickinson nor I have any birds, which is also to say we have no lyrics—at least not as her critics know and love them. It is not possible to *have* lyrics in theory without having *made* them historically, as Jackson observes (11). Her argument for lyric reading can itself be read as an inverse (or perverse) form of generic skepticism in contrast to the usual practice of reading lyric. If reading lyric assumes that the generic object precedes the act and therefore insists that we (intuitively) *know* a lyric when we see it, Jackson proposes instead that believing is seeing—"to be lyric is to be read as lyric" (6)—so that we can see or read a lyric only when we (already) know it. Jackson opens *Dickinson's Misery* with a hypothetical scenario about the discovery of the fascicles after Dickinson's death, in the moment of seeing (or believing) that these are lyric objects: "Suppose that you recognize the twined pages as sets of *poems* . . . what remains, you decide, must be published" (1). Although we cannot overturn a century and a half of hypothesis-turned-hypostasis to return Dickinson's written remains to a time before reading lyric, lyric reading can account for this history while holding open the theoretical possibility of the genre as a "virtual object" (10). To read an object lyrically is to suspend the critical beliefs and historicize the editorial decisions that would turn abstractions into realities, translate bound pages into reading editions, and bring lyric to life. Below, I dissect recent critical readings of "Split the Lark" for their assumptions and hypostatizations, which vivify their object by positing an invisible spirit that transcends and can therefore *return* (or exist again)—the apotheosis of a vitalist lyric faith.[3] Then I propose a materialist but hypothetical reading with the help of Daniel Tiffany's theory of lyric substance. My aim is neither to one-up critical vivification with concrete historical artifacts nor to assume the existence of real artifacts through a naïve materialism or faith in brute matter (which Tiffany would caution against). Rather, I will conduct a kind of experiment that asks what critics see, know, and believe—visible or invisible, historical or conceptual—and what we might miss when we look only for certain kinds of birds or lyrics or fear their fatal split.

"Split the Lark" hasn't spilled as much scholarly ink as some of Dickinson's other lines, but its opening imagery has inspired persistent speculation:

> Split the Lark - and
> you'll find the Music -
> Bulb after Bulb, in
> Silver rolled -

What bulbs? What silvery roll? Where *will* we find the Music? Critics tend to read this imagery as a kind of vital and feminine potential hidden both within Dickinson and in the lyric genre. Helen Vendler suggests that the silver bulbs are an image of "the potential of poetry in [Dickinson] herself . . . like eggs in the ovary . . . or flower bulbs ready to unfold," whose "future notes" are "flutelike, 'in Silver rolled'" (367–368). The same bulbs have evoked everything from pearls (Nyberg, "Translating," 113) to the goose's golden eggs (Wolff, *Emily Dickinson*, 362), from "life potential lovingly wrapped in the silver of song" (Leiter, *Critical Companion*, 176) to Kamilla Denman's holistic simile: "Like volcanic lava or the waters of a flood, the music of the bird resides within, and its expression is an overflowing of inner essence" (34).[4] These interpretations also echo with a persistent set of inherited lyric discourses; the inner unfoldings and affective overflows that they read into Dickinson's opening lines hearken back to John Stuart Mill's 1833 definition of poetry as "overheard" or "feeling confessing itself to itself, in moments of solitude" (12). For Mill, though poetry may (in fact, must) solicit an audience, its speaker must equally pretend unconsciousness of that audience and uphold the presumed privacy of reflexive address. His simultaneous isolation and idealization of poetry has become a touchstone for lyric's separation from any context, including the economy of its circulation in print (on "hot-pressed paper," as Mill has it). By positing poetry as affective soliloquy and adopting what Jackson characterizes as Mill's "will to lyricize" without taking up his somewhat defensive (because indefensible) assertions about the essential difference between poetry and other genres (*Dickinson's Misery*, 130), subsequent critics are able to pretend that lyric floats free from the millstone of history and the complexities of generic address (132).

Secure in their belief in the power of affective overflow, few close readings venture on to the subsequent lines, which first qualify the quality and volume of the overflow, and then suggest the potentially unnatural repetition or preservation of this Flood, to be loosed by "you" at will:

Scantily dealt to the
Summer Morning
Saved for your Ear, when
Lutes be old.
          Loose the Flood -
you shall find it patent -
Gush after Gush,
reserved for you -

Averting their gaze from such gory allegory, which seems to swap the overflowing of inner Music for bloody innards, critics focus instead on the solid ground of religious allusion in the final lines:

> Scarlet Experiment!
> Sceptic Thomas!
> Now - do you doubt
> that your Bird was
> true?[5]

From this vantage point, "Split the Lark" is generally read as a straightforward send-up of the empirical science that would split birds, atoms, or hairs (whatever the consequences). As Lennart Nyberg observes, "[the poem] is about the skeptic's need to find physical evidence of the invisible" (113). This clash between faith and doubt hinges on the figure of Doubting Thomas, who refuses to believe the resurrection without "knowing" the evidence of Christ's wounds, and insists on seeing the marks on his hands and placing his hand in Christ's side. Christ invites him to do so but admonishes him not to be "faithless, but believing" (John 20:24–29). For Christopher Benfey—and, he believes, for Dickinson as well—"Thomas's doubt and his demand for proof are as faithless and murderous as the demands of the crucifiers" (94). The result of the skeptic's "scarlet experiment" is not therefore the discovery of the Music but "a dead bird" (94).

If the clash in "Split the Lark" is between preserving "birds" or killing them to make visible the inner evidence of their Music (rather than overhearing its invisible essence), critics very much want to be "true": to come down on the side of faith in lyric and avian life rather than having blood on their hands as unbelievers. As much is audible in the warning that Susan Stewart reads into the poem:

> Taking apart something in order to understand it inevitably raises the possibility of killing the whole. The music is a manifestation of the lark's entire being and to try to find it through dissection would be to discover the hidden affinity between the bird's music and its blood; to spill it by artificial means would be to drain the body of its life. (158–159)

Here the lark's being is "manifested" not only in and as its music but in its blood and therefore as if in its life. The "hidden affinity" that hypostatizes music as blood in the final "it" ("to spill it") is followed immediately by their mutual vivification in and as if "life" itself. This critical logic moves beyond lyric's immanence in a particular manifestation to its invisible and transcendent presence. For Judy Jo Small, "The song . . . is impalpable, not contained in the physical mechanism of the bird's body, and it cannot be separated from the secret of its life" (38), and for Carolyn Cooley, it's "futile" to seek to extract the Music from "the *living* lark . . . whose nature it is to sing and whose life is reflected in song" (45). The effect of this hypostatization is not only to vivify a genre; it also renders that life apparently "natural" rather than artificial (that is, biological and essential rather than mechanical or historical). On the whole, critics come down in favor of the transcendent vitality of Dickinson's corpus and so of lyric itself. As Denman has it, "Split the Lark" makes "the inorganic matter of language into living poetry" (42), or, in Stewart's assessment, it reveals the "interdependence of immanence and transcendence in any artwork" (159). For Stewart, while the "life of the work is embodied and must precede transcendence," transcendence arises nonetheless from these conditions: "the 'music' of organic wholes in nature and in art cannot be traced to any particular and cannot be accounted for by a knowledge of its merely material source" (159). Here "music" is the vitalist force or spark, the élan that vivifies particular wholes and survives them. The faithful critic therefore believes in lyric's continued generic life without asking after its origins, its inscription, or its insides—without needing to know, see, or touch the material source of its secret and invisible song. To seek this "merely material source," to discover its working parts or expose its physical mechanisms (or the history of its particular manifestations), would risk the genre's veneer of spontaneous overflow, its affective immediacy (the assumption that we know a lyric when we *feel* one), its reflexive privacy (despite its circulation), and so its presumed life. The concept of lyric as literary object at work in these readings assumes that any "manifestation" is separate and separable from its historical or material artifact, positing an invisible lyric spirit that survives beyond mere stuff or biological mechanisms. The deeper article of faith at work is not just that there's something to be killed (or murdered) by skepticism but that something of the lyric remains to be saved—and that it might well save us.

These readings are the epitome of the lark as lyric allegory, and their tendency to equate the poet with an avian avatar is our inheritance from a familiar romantic discourse around birdsong as the ideal figure of lyric expression.[6] Certainly Dickinson's target in "Split the Lark" is no biological bird; by the mid–nineteenth century, Jackson tells us, "a bluebird's (or a nightingale's or a sky-lark's or a bobolink's or a darkling thrush's) tune would always already have been . . . a lyric poem" (*Dickinson's Misery*, 27). Shelley's skylark was never a biological bird either, but a "blithe Spirit" that pours out its "full heart/In profuse strains of unpremeditated art" ("To a

Skylark" ll. 4–5). No scanty dealings in this full-throated song, which the poet cannot hope to imitate except perhaps as starling or parrot. On Jackson's reading, "Split the Lark" eviscerates the "oxymoron" of this "embodied abstraction" by its "sadomasochistic" performance (*Dickinson's Misery*, 187, 189). Where Shelley's bird bleeds only music, Dickinson's lark bleeds red. In this sense, the warning in "Split the Lark" is not that we must have faith in the invisible and transcendent source of "Music" or lyric beyond the body of the bird (or beyond the reach of the poet). Rather, the target is the hypostatization of lyric itself; as Jackson has it, "lyric—or, specifically, the reading of it as an indexical cultural identity—may be the object of Dickinson's perverse anatomy lesson" (*Dickinson's Misery*, 187). What if the imperative or invitation that opens "Split the Lark" comes not from a member of the faith but from an unbeliever who is also not a skeptic? That is, if the skeptical insist on seeing evidence in order to believe, and the faithful believe without needing proof, "Split the Lark" may split the difference by staging the act of lyric reading itself.

Franklin's #402, "To hear an Oriole sing," already stages what we might call lyric *hearing* as a dialogue with "the skeptic":

> To hear an Oriole sing
>
> May be a common thing—
> Or only a divine.
>
> It is not of the Bird
> Who sings the same, unheard,
> As unto Crowd—
>
> The Fashion of the Ear
> Attireth that it hear
> In Dun, or fair—
>
> So whether it be Rune,
> Or whether it be none
> Is of within.
>
> The "Tune is in the Tree -"
> The Skeptic - showeth me—
> "No Sir! In Thee!"

As in Jackson's "to be lyric is to be read as lyric," to hear this oriole is to hear not "of the Bird" but of "Thee." This inverts the usual (human) lyric subject (because "Thee's" are the wrong insides to be the subject of Mill's lyric expression), while the bird's inhuman lyricism remains romantically unaffected by any audience, singing "the same, unheard, / as unto crowd." If the skeptic's birds are in the tree and his tunes are in the bird's innards, and if the romantic poet's darkling bird is invisible in the bush and its song is inimitable (but certainly worthy of a poem), in the above readings of "Split the Lark," the bird is lyric (poet) and her Music is transcendent and vital. "Split the Lark" takes these assumptions and inversions firmly in hand, exploding the (living) bird and the genre that critics and skeptics thought they had or knew. In response, and in an attempt to save the lyric from *their* insides out (that is, to reverse again the inverse skepticism of lyric reading), critics tend to take themselves and the act of reading lyric to be Dickinson's target: "By reading the work of the poet addressing us we are destroying her gift—splitting open the source of the private and privative music" (Raymond, *Witnessing Sadism*, 107). Potentially guilty of parsing images and interrogating every comma and dash, they seek to avoid the self-fulfilling prophecy by which "logical analysis quickly becomes murderous dissection" (Small, *Positive as Sound*, 38), effectively splitting off the historical and material practices of writing, editing, and reading from the immaterial "source" and ethereal life of lyric. Though the skeptic and the scientist have no birds—or worse, have murdered them—the faithful want to have their bird and keep lyric alive, too.

Heedless of critical caution or parodic warnings, and no doubt conscious of the chorus of romantic birds on the line, Legault's rendition of "Split the Lark" shoots from the hip: "By dissecting a bird you can locate its vocal chords" (*Reader*, 121). We might hypothesize that the speaker of this sterile and seemingly scientific statement is none other than the addressee of "Split the Lark": the skeptic who has faith in physical evidence alone. But if this skeptic's one-liner reads like textbook instructions, that text gives a false account of the origin and anatomy of birdsong. Birds (in fact) lack the vocal chords and larynx of the human throat; instead, they produce sound from the lower end of the windpipe, through the syrinx. Precisely because of this false factoid, Legault's performance rings doubly "patent" or true to its (invisible) source by one-upping Dickinson's send-up. While "Split the Lark" calls out the romantic abstraction of disembodied lyric birds by taking it literally, Legault's sleight-of-hand replaces an invisible and abstract ideal (Music) with an invisible because nonexistent body part—one that we couldn't see for looking, skeptic or scientist though we may be. Perhaps, in this mashup of human and avian organs, Legault's bird has swallowed the critic-reader, or vice versa; insofar as its innards are now ours, a bird with human vocal chords seems an apt figure for the projection performed by "Thee" in "To hear an Oriole sing."[7] Rather than its hidden music living in "me" (Mill's lyric subject) or "the tree" (say, its material manifestation), this song—whether heard as "common thing" or "divine," "Rune" or "none"—lives only in the "Fashion" of its hearer's ear. In other words, the skepticism of lyric hearing overtakes the faith of lyric hearing (so that hearing is believing rather than believing before hearing).

In what follows, I want to combine Legault's pseudoscientific substitutions and Jackson's anatomy of lyricization with Daniel Tiffany's "lyric substance" to pose a hypothetical reading of "Split the Lark." Jackson is no doubt correct in supposing that one of the things you find when you split a lark is poetry (*Dickinson's Misery*, 187). I would extend that theoretical and historical reading by nominating another (virtual) object as the thing you find. What if Dickinson's lark is nothing less—or more—than a music box, that nineteenth-century version of the serinette or bird-organ? Split this lark and you'll find neither blood nor beautiful Music-with-a-capital-M, neither generic life nor allegorical silver bulbs, but silver rolls and mechanical bells. If Dickinson's lark is not just the evisceration of a disembodied ideal but the substitution of that vivified ideal for a mechanical object, what effect would it have on the lyric assumptions that accompany critical readings of "Split the Lark"? If its presumed bird becomes instead a hypothetical bird-organ, "Split the Lark" goes one further than staging its own hypostatization; it also stages its own death, its visible machine proving fatal to the critical faith in lyric life.

First, a little history to accompany our hypothetical mechanical lark. Invented in France in 1730, the serinette was named for the canary (*serin*) and consists of a miniature barrel organ housed in a wooden case. To play it, you turn a crank on the front of the case, which pumps the bellows to supply air to the tin pipes and turns the wooden barrel on which metal pins or staples are plucked by the wooden pegs mounted above the cylinder. This roll—or the steel roll of the later music box—is where you'll find the Music:

> Split the Lark - and
> you'll find the Music -
> Bulb after Bulb, in
> Silver rolled -

The serinette was the first means of recording music and the newest gadget of the day (Birkhead, *A Brand New Bird*, 65), although mechanical musical instruments—that is, instruments with self-playing capacity—have existed "almost as long" as "music itself" (Ord-Hume, *Musical Box*, 23).

Suppose, then, that the music you've found is a recording, "scantily dealt" from the thin cylinder as you slowly turn the crank, or in gushes if you turn faster or unevenly:

> Saved for your Ear, when
> Lutes be old.
> Loose the Flood -
> you shall find it patent -
> Gush after Gush,
> reserved for you -

Not only does this hypothesis shift the sense of the possessive "saved" and "reserved" from the religious or romantic connotations assumed in the readings above; it doubles the critical horror over the fatal revelation of the "physical mechanism" of any bird's tune, because the box's abstract and transcendent music comes not from some invisible physical organ within a living bird (which would be bad enough) but from a bird-organ—a machine. This machine strikes an equally fatal blow to the lyric subject assumed in any "will to lyricize." First, the generic thing we thought we saw (or heard) no longer emerges in "utter unconsciousness of a listener" to express subjective "feeling confessing itself to itself" in solitude; second, and worse even than Mill's eloquence, which is "feeling pouring itself out to other minds," this music pours out without feeling or direction (12). It begins not when the speaker is moved, and regardless of whether its listeners are, but whenever someone moves the crank or opens the lid, and it can be replayed when other sources of music "be old," whether Lute or lark. We shall find this music "patent" in more than one sense, too: Its mechanism is evident to the eye when the box is open, and its music is poured indiscriminately into any ear. More, this mechanical performance shifts from the perpetually present and suspended "now" of lyric's assumed temporality (in the opening lines) to a perpetually available music on demand, often with more than one tune; the bird-organ's cylinders could be exchanged to play different songs.

In addition to outlasting any live performance, the machine's music was intended to move not humans but birds as the primary audience of its eloquence. Initially, particular species were selected for their ability to mimic songs, and nightingales or larks were used to train domestic caged birds. These avian trainers were later replaced by "more reliable" (because mechanical) instruments, including flageolets and the piccolo-like tones of the serinette (Gessert, *Green Light*, 16; Birkhead, *A Brand New Bird*, 62–64). The instruments could be used to mimic natural birdsong by producing "continuous rippling notes" (Gessert, *Green Light*, 16), but most appear to have played human airs, whether popular—such as Handel's *Rinaldo* and "The Happy Clown" from *The Beggar's Opera*—or composed for the purpose and incorporating "trills and other birdlike sounds" (Birkhead, *A Brand New Bird*, 63). Whether playing "natural" or human airs, the serinette's repertoire could be modeled to suit the range and pitch of different species; the most common was the canary, but models were also available to train blackbirds, bullfinches, and even parrots. Whatever their pitch, however, it's not clear that these machines were capable of producing or inducing the desired effects in their pupils—that is, of teaching the bird to imitate tunes that in turn mimicked its vocal range, if not its song. Birkhead observes that the resulting performances "often left much to be desired," citing Carl Philipp Emanuel Bach's exhortation to his (human) pupils to "'play from the soul, not like a trained bird!'" (65). From this perspective, Dickinson's closing question is less about religious faith or human loyalty and more about whether the singing bird, mechanical or caged, was accurate and in tune:[8]

> Now - do you doubt
> that your Bird was
> true?

Fidelity in this musical sense was an important question for later larklike technologies, too. By the time Dickinson wrote "Split the Lark" in 1865, the serinette and its caged pupil had been overshadowed by the rise of the music box proper. These "musical boxes" could be played by crank, as bird-organs were, or by winding up the spring with a key or a handle; some even used pull-cords to wind the spring that plucked the pins on a cylinder with the teeth of a steel comb (Ord-Hume, *Musical Box*, 102–109).[9] The use of the steel comb mechanism was first documented in the 1790s in Switzerland as a novelty to accompany timepieces (where it remained hidden in their bases, as in cuckoo clocks).[10] By the 1820s, music boxes had become solo instruments in their own right, and their intricate mechanisms were patently visible to their audiences, displayed below glass in increasingly ornate cases (41). As the technology advanced, drums, castanets, and bells—some even struck with bird-shaped hammers—were added to accompany the tunes on the cylinder (219, pl. 112–113). Just as the cylinder was hidden in the base of the cuckoo clocks it accompanied, these percussive accompaniments were initially hidden in the bedplate of the box, but by midcentury, their bells and whistles had been made visible and were marketed in music boxes with names like *Visible Bells* and *Bells in View* (217, pl. 106–107). Music boxes became more widely available in the last quarter of the century in America and in Europe, when disk boxes began to overtake the earlier cylinder style. Ord-Hume attributes the eventual decline of the technology to the outbreak of the First World War and the invention of the phonograph, "an instrument which could reproduce any number of sounds including the human voice in all its inimitable nuances"—despite the music box having what he calls an exclusively and inimitably mechanical "voice of its own" (30). Ord-Hume's attribution seems symptomatic of the tension between human and inhuman lyricism via the figure or body of the bird: a recording technology invented to train birds to mimic human airs that fails because it cannot reproduce the human voice. This music box never had a voice box. Like Legault's bird, the bird-organ has no vocal chords, and like Dickinson's lark, what dies when we replace the hypostatized lark with the serinette is neither bird nor lyric but their vivification.

To return to the bird at hand: By the time Dickinson wrote "Split the Lark" in 1865, there may not have been a cylinder music box in every New England household, although they would have been more common in wealthier ones like hers. It therefore seems probable that she knew of them, whether she had one in hand or not. Beyond this assertion, my music box proposition and the lyric reading that follows from it must remain hypothetical. Kind of like birds, neither Dickinson nor I have any mechanical musical instruments. I can find no archival evidence or direct textual reference to a music box, no proof that Dickinson or her family owned one.[11] But we need not maintain blind faith in an invisible or nonexistent music box any more than we need insist on material evidence that Dickinson must have seen or heard such a box in order to write "Split the Lark"—or in order to reread it as if she might have. In other words, an unsubstantiated music box can make a difference in our hypothetical reading of a hypostatized object like the lyric, as demonstrated by Daniel Tiffany's *Toy Medium: Materialism and Modern Lyric*. Where Jackson sets the historical and material record against the lyricization of Dickinson's printed matter in order to reread it, Tiffany sets the historical and philosophical record against the naïve materialism of contemporary cultural studies, which would conflate materialism with realism (1–2). Tiffany complicates this too-simple "equation of materiality and visibility" by collecting a dazzling range of texts and objects—from toys and automata to meteorological phenomena—to define what he calls lyric substance.[12] Tiffany's lyric substance deals in a "material occult" (3) that might "make the intangible tangible" (5), thereby dissecting critical faith in visible matter as the only "real" and offering a new model for materialist criticism (32).

Tiffany's sleight of hand is to use modern lyric as the genre through which to convince the naïve critic that there's more to matter than meets the eye. He opens his attempt with a thought experiment that would appeal to lyric skeptics and faithful critics alike: What might lyric substance share with both "the materials of poetry itself—the substance of words and pictures"—and "scientific explanations of matter" (15)? Fittingly, the model he introduces to explore this hypothesis is Yeats's mechanical bird in the Byzantium poems. Like "Split the Lark," Yeats's automaton too revises the romantic bird, but unlike Dickinson's, it retains its immateriality. For Tiffany, the golden bird's "sensual music" in "Sailing to Byzantium" signifies "both the lyric medium and corporeal fate of the poet" and therefore illustrates the "dual nature of lyric substance" (19). In both poems, for Tiffany, Yeats's depiction of the soul as a mechanical toy means that "the lyric poem itself—its body—is the primordial image of the material soul" (27). Rather than opposing the soul to corporeal existence, this soul is manufactured: an artifact rather than a property of the mind (19). If the music box hypothesis about "Split the Lark" rings true, then Tiffany and Dickinson might agree; the material soul becomes a singing automaton or a bird-organ rather than a hypostatization of lyric's so-called life. In this reading, Tiffany successfully sidesteps the assumed vitalism that usually attends more-than-material arguments in favor of the lyric, which operate on the vitalist assumption that avian—which is to say lyric—life transcends its corporeal manifestation. In fact, for the faithful critics critiqued above, the lark doesn't die because its material body is dissected; crucially, it dies because "you" (the skeptic who failed to believe without seeing) sought its musical soul, the immaterial and invisible but essential—read vital—quality that depends on but preexists (and potentially survives) its physical mechanisms or corporeal source. But here the music box splits from the mechanical bird over their physical mechanisms and (in)visible insides. On Tiffany's reading, the automaton or the mechanical toy is significant because its mechanism is "*internal*, hidden from view"; unlike other pictures, "it possesses an inside, an inscrutable interior" that confirms the hypothetical method of mechanical philosophy (53). Its invisible mechanism (with observable effects) thereby confirms the probable existence of the invisible substance of things (like atoms). The skeptic proper might not believe in this method, but the scientist could, and so should the humanist critic (if we believe in Tiffany's lyric substance). By contrast, it is more significant for "Split the Lark" that the mechanism of our hypothetical music box is patently on display (by the end of the century, anyway), making it a model not for the hypotheses of mechanical philosophy but for imploding the hypostatization of lyric in and as invisible avian music. If the lark's mechanism remained hidden, as it is in Yeats's mechanical bird, its eviscerating effect on critical faith might not be as pronounced, and lyric vitalism might live another day.

In this vein, the visible and mechanical music of "Split the Lark" might also serve to split the "sensual music" carried over from Yeats's bird to Tiffany's hypothesis or at least make visible the continued generic assumptions on which they rest. In reading "Sailing to Byzantium," Tiffany triangulates—perhaps conflates—lyric and "corporeal fate" with "the point where the automaton becomes a *living* picture" (21): In other words, "the mechanical bird comes to life when it sings" (19), and its "sensual music" is at once the cycle of biological life *and* the song produced by the poet. The shared secret of avian automata and pictures in general lies in their *animation*, which Tiffany reads as a broader symptom of the uncertain "*materiality* of images" (23). Given the resonance of his reading of "sensual music" with the critical readings of the lark's music above, Tiffany's recommendation that we "refer" the problem—or the fetish—of "live pictures" to the lyric principle of a material soul seems merely to defer or invert that problem by maintaining the assumption that lyric speaks uniquely or primarily (even primordially) to an equally hypostatized concept of "'life' (however that may be conceived)" (22). More, since the publication of *Toy Medium* in 2000, the "explosion of materialist criticism" that Tiffany identified at the time has changed direction or merged with the recent "explosion" of vital materialism, which itself appears to take an unexamined lyrical turn in its emphasis on animation (as in Jane Bennett's *Vibrant Matter*) (31). It would seem that life has overtaken matter as the current "problem," and the stakes and implications are just as high (32); this nouveau vitalism is often accompanied by the brute materialism that Tiffany critiques, but it shares and therefore makes visible the continued assumption of the genre's priority and its apparently innate ability to speak to (human) life.

While Tiffany's *Toy Medium* allows me (not) to substantiate my music box hypothesis, then, we might also allow "Split the Lark" to prevent his careful hypotheses from advancing too far into assumptions or necessities. That there are substances in the world that humans can't see comes as no surprise, but that Tiffany asks us to "acknowledge that materialism possesses an inherently lyrical dimension" may (290). Similarly, we can grant that scientific knowledge too relies on mediation or representation—that is, on pictures—but must that reliance be inherently "lyrical"? And must it be "indispensable to the appearance of the phenomenon it depicts" (291)? Or do these claims invest too much substance or power in an adjective ("lyrical") derived from a virtual genre? Tiffany notes that proposing "that lyric poetry might serve as a model for materialist criticism . . . possesses a certain charm" (290), but he doesn't say for whom; we might hypothesize that the idea holds more charm for (faithful and enchanted) lyric critics who wish for greater visibility or substance for their "toy" genre than it does for scientists. While his call for greater lyricism may counteract the naïve materialism he identifies in cultural studies, and in some cases in historical criticism, his interest in using poetry as a model organism (or toy) lies in regaining its privileged position as the "picture" or analogy that "matters" most—or that should, both to literary and cultural studies and in "public debate" (6, 9).[13] Tiffany sustains a residual faith in lyric as first among genres, as "primordial" or primary, calling on Aristotle, Lessing, Kant, and Heidegger to articulate its position at the "top of the hierarchy of the fine arts," and thereby making a case for lyric's immaterial body as a model for the "disembodied pictures of contemporary media" (27). But as Jackson would remind us, if we never had lyric in the first place (at least not without first believing in it), and if it was made historically, how can it (pre)exist or be found without first hypothesizing—and then hypostasizing—that existence? Tiffany certainly doesn't dismiss historical criticism, but he suggests that "its claims to authenticity, as against a more philosophical approach to the object, should be tempered in light of the inescapably lyrical—and even phantasmagorical—nature of materiality" (289). Without dismissing his "more philosophical" criticism, we can still temper its claims for the priority and necessity of lyric in light of the inescapably historical, as well as phantasmagorical, "nature" of the genre. In this light, referring the problem of animation and uncertain materiality to *lyric* as aesthetic index comes to look like a symptom of critical faith by other means.

I want to conclude with a return to lyric criticism beyond "Split the Lark," using a recent example to make visible the (virtual) object that faithful critics aim to save when they ascribe philosophical or aesthetic priority to the genre. That object is none other than humankind. In splitting the lark, Dickinson's lines equally eviscerate the persistent humanism that subtends vitalist critical readings of the lyric as more than material, as if the genre was at once the height and heart of (human) linguistic expression. Eavan Boland's article on "The Serinette Principle: The Lyric in Contemporary Poetry" revives the Millsian will to lyricize and makes explicit the will to save the genre—and thereby ourselves—by hypostatizing the lyric as serinette and recuperating the machine as human organ.[14] She opens with her own discovery of the history of the instrument, used "for the express purpose of teaching caged birds to sing" (312). In keeping with the lyric and avian tradition we've inherited, Boland assumes that the song being taught to the caged bird is its own—that is, that the serinette was used to teach domesticated birds to sing the "natural" songs they might have sung in the wild, rather than teaching them to imitate a human air or a popular tune. Following from this assumption, Boland's hypothetical serinette figures an authentic and innate "lyric impulse" that "reaches out to a perceptive area that has fallen silent" both in the caged bird and in "us" (312). For Boland, the lyric genre is our serinette, successfully making patent—and true—a previously imperceptible but no less innate "internal power within us" (312). The lyric's ultimate aim is to "charm and to heal our sense of time and terror" (313),[15] and the genre is therefore able to restore us to our origin, beyond the modern age or cage of time (313): "This is the true lyric moment. We no longer hear the serinette. What we are listening to is our own song" (327). For Boland, we rediscover our immaterial soul "by hearing its external imitation" (312), and when "true," that imitative medium again becomes invisible, vanishing in the transhistorical song intrinsic to the human species.[16] However its form or manifestation may change over time, then, "the lyric is inscribed on the history of the human race and the origin of language," and its essence remains to be resurrected (328).[17] No merely mimetic technologies or manufactured souls here, and none of the nuance of the oriole's inverted song; as lyric allegory, the romantic bird's hardly inhuman lyricism becomes a reflection of humanity's inner song, just as Boland assumes that the serinette's tune reflected its bird's inner song (and so its life beyond the cage). Where Legault's bird may have swallowed the critic to turn its song and our insides out (or into bird parts), and where the bird-organ disgorges only mechanical bulbs and tunes-on-demand from all-too-patent parts, Boland begins from the serinette principle to return to an all-too-human lyric impulse. For Boland as for other faithful critics, the genre's adopted avian life must both precede and exceed its technological instantiation and cannot be reduced to its material (re)production or historical context. More, we must believe in this transhistorical avian and lyric life as if it at once indexes and affirms *our* species' inner song and continued life: "Poetry lies within us. Its first tentative sounds must be stored in us somewhere" (329), and "we" the species will instinctively know (and feel) it when we hear or see it. Here the human body become the serinette—the machine in which the song is stored—and our soul or being lives on in the lyric that transcends both its temporal inscription and its temporary dissection. Like readers of "Split the Lark" who see flower bulbs and ovaries instead of metal pins and silver cylinders, Boland neatly recuperates the bird-organ as human organ, perhaps our heart or an inner voice, a lyric potential lying deep "within us." But as Jackson suggests, rather than parroting the critical faith that believes in "poetry as lyric and lyric as poetry that might save us" ("The Beginnings of Poetry"), we might instead begin to historicize this vitalist faith in origins (and in more than mechanical organs) and to hypothesize rather than hypostasize lyric as much as life. Dickinson's music box offers one way to split such persistent faith in the lark with starling skepticism.

NOTES

1. See Socarides's "For Emily, Wherever I May Find Her" for an overview of responses to Legault's collection that range from outrage to delight. The collection as a whole exhibits his magpie-like tendency to collect Dickinsonesque imagery, objects, and obsessions, drawn eclectically from all possible contexts (biographical and historical, critical or anachronistic), shot through with a variety of birds (bobolinks in particular), flowers, zombies, and death, and crossed with regular references to books, sex, and "Sue" (Susan Gilbert Dickinson, Emily's sister-in-law).

2. Legault: "For any American poet, Emily Dickinson is sort of a monolith. There's no way around/over/under—you have to go through it. To me, translating Dickinson seemed as inevitable as a contemporary musician covering Bob Dylan" ("Emily Dickinson"). See also Jackson on readers' persistent "preoccupation" with the "exemplary American character" of Dickinson's work (*Dickinson's Misery*, 11).

3. My examples come from the 1980s to the present and are intended to be indicative rather than exhaustive.

4. In contrast to these politely allegorical readings of (problematically) gendered lyric fertility, William Shurr hypothesizes historical gore: "if" Dickinson suffered a "painful abortion . . . then" "Split the Lark" "garishly asserts that her love for the 'you' of the fascicles was proven 'Gush after Gush' in a 'Scarlet Experiment'" (181). And for Leiter, it's hard not to see "orgasm in the image of the silver-rolled bulbs, the flood of virginal blood and/or passion" (176).

5. Franklin, *The Poems*, 905; the line breaks in this version are reproduced from Jackson, *Dickinson's Misery* (185–187 and the ms. pictured in Fig. 28).

6. Whether Dickinson herself is figured as (too) avian or (all-too) human depends, of course, on who is reading and when (Jackson, *Dickinson's Misery*, 128). Critics of "Split the Lark" are largely split between the romantic figuring of birdsong as a form of overheard and "inhuman lyricism" and the later New Critical reception of Dickinson's texts as the isolated (human) voice of subjective expression (128).

7. It's important to note that this inversion does not make "our" innards universal; whereas the lyric poet-lark's "I" and its addressee blend quickly into a universal and transhistorical human "we" (Jackson, *Dickinson's Misery*, 129), Dickinson's oriole insists on subjective rather than shared experience, and the lark's irony splits a set of assumptions particular to a specific historical period (which change with their later critical adoption).

8. The fascination with training birds to imitate tunes is encapsulated in Jean-Eugène Robert-Houdin's 1843 *La serinette*: a pair of automata that consisted of a mechanical lady playing a serinette to a mechanical bird, which would then sing back the tune that the serinette had just played and whose initially imperfect imitations would improve in accuracy on subsequent playbacks, until finally the imitation was "true" to the serinette's original tune—itself already an imitation (Stamp, "A Brief History").

9. Arthur Ord-Hume is quite firm: in order to be called a "musical box," the sound must be produced by steel teeth plucked in sequence (*Musical Box*, 40). The term "music box" refers, for Ord-Hume, to a box that holds sheet music; a *musical* box is one that *produces* music. I use the more colloquial music box to refer to musical boxes throughout.

10. See Ord-Hume, *Musical Box*, chap. 2, for a careful discussion of the exact chronology of its use and invention, which he believes to be earlier than generally assumed—as he points out, the first *documented* use is not to be equated with first use (46).

11. Many thanks to Sandra Runzo and Jane Wald for searching concordances and inventories for any reference to a music box.

12. Tiffany's lyric substance is both a concept and a property that he uses to model the ways in which matter can be visible or invisible, tangible or intangible; in his words, it is a "doctrine of corporeality" proper to the genre and potentially possessed by other objects and bodies (*Toy Medium*, 15). It is therefore distinct from hypostatization proper, which would treat an abstract concept as if it had real substance.

13. This despite his critique of the "ethical or epistemological priority frequently assigned to materialist criticism" (Tiffany, *Toy Medium*, 2); W. J. T. Mitchell makes the same case in *What Do Pictures Want?*

14. Boland concedes in conclusion that her serinette gimmick is "a construct," perhaps a hypostatization ("The Serinette Principle," 329).

15. This claim resonates with Stewart's *Poetry and the Fate of the Senses*, which holds poetry up as a kind of shared sixth sense, a transcendent Kantian "pure form" of intuition: "As first-person expression in measured language, lyric poetry lends significant—that is, shared and memorable—form to the inner consciousness that is time itself" (42).

16. Again as in Stewart: "the ways in which poetic forms are immortal is in truth very much like the continuance of human life by physical and social means" (*Poetry and the Fate of the Senses*, 331).

17. As do most faithful critics, Boland argues for the genre's vital power to reinvent itself across cultures and times, regardless of its technological or material instantiation: "We are looking at a form that has fitted itself, on different occasions, to the lyre, to the lute, to the harp; to the small kingdom and the lost tribe. We could find it, if we searched, on the Pyramid texts of Egypt, under the battlements in Picardy; on the back roads of Ireland in a defeated language" ("The Serinette Principle," 313).

WORKS CITED

Benfey, Christopher. *Emily Dickinson and the Problem of Others*. Amherst: University of Massachusetts Press, 1984.

Bennett, Jane. *Vibrant Matter: A Political Ecology of Things*. Durham, N.C.: Duke University Press, 2010.

Birkhead, T. R. *A Brand New Bird: How Two Amateur Scientists Created the First Genetically Engineered Animal*. New York: Basic Books, 2003.

Boland, Eavan. "The Serinette Principle: The Lyric in Contemporary Poetry." In *Parnassus: Twenty Years of* Poetry in Review, ed. Herbert Leibowitz, 311–329. Ann Arbor: University of Michigan Press, 1994.

Cooley, Carolyn Lindley. *The Music of Emily Dickinson's Poems and Letters: A Study of Imagery and Form*. Jefferson, N.C.: McFarland, 2003.

Denman, Kamilla. "Emily Dickinson's Volcanic Punctuation." *The Emily Dickinson Journal* 2, no. 1 (Spring 1993): 22–46.

———. "Review of *Poetry and the Fate of the Senses*, by Susan Stewart." *Modern Philology* 101, no. 3 (February 2004): 506–509.

Franklin, R. W., ed. *The Poems of Emily Dickinson*. Reading ed. Cambridge, Mass.: Belknap, 1999.

Gessert, George. *Green Light: Toward an Art of Evolution*. Cambridge, Mass.: MIT Press, 2010.

Higginson, Thomas Wentworth. "Emily Dickinson's Letters." *Atlantic*, October 1891.

Jackson, Virginia. "The Beginnings of Poetry," *Toronto Review of Books*. Podcast. November 18, 2011. http://www.torontoreviewofbooks.com/2011/11/trb-podcast-the-beginnings-of-poetry-virginia-jackson-at-the-university-of-toronto/.

———. *Dickinson's Misery: A Theory of Lyric Reading*. Princeton, N.J.: Princeton University Press, 2005.

———. "Who Reads Poetry?" *PMLA* 123, no. 1 (January 2008): 181–187.

Legault, Paul. *The Emily Dickinson Reader: An English-to-English Translation of Emily Dickinson's Complete Poems.* San Francisco: McSweeney's, 2012.

———. "Emily Dickinson's a Pisser: Talking to Poet-Translator Paul Legault [Interview with Mark Asch]." *The L Magazine* (September 4, 2012).

Leiter, Sharon. *Critical Companion to Emily Dickinson: A Literary Reference to Her Life and Work.* New York: Facts on File, 2007.

McLane, Maureen. *Romanticism and the Human Sciences.* Cambridge: Cambridge University Press, 2000.

Mill, John Stuart. "What Is Poetry?" In *Essays on Poetry*, ed. F. Parvin Sharpless, 3–22. Columbia: University of South Carolina Press, 1976.

Mitchell, W. J. T. *What Do Pictures Want? The Lives and Loves of Images.* Chicago: University of Chicago Press, 2005.

Nyberg, Lennart. "Translating 'Split the Lark' Into Swedish." *The Emily Dickinson Journal* 6, no. 2 (Fall 1997): 112–117.

Ord-Hume, Arthur W. J. G. *Musical Box: A History and Collector's Guide.* Boston: George Allen, 1980.

Raymond, Claire. *Witnessing Sadism in Texts of the American South: Women, Specularity, and the Poetics of Subjectivity.* Farnham: Ashgate, 2014.

Shelley, Percy Bysshe. "To a Skylark." In *Norton Anthology of Poetry*, 3rd ed., 624–626. London: Norton, 1983.

Shurr, William. *The Marriage of Emily Dickinson: A Study of the Fascicles.* Lexington: University Press of Kentucky, 1983.

Small, Judy Jo. *Positive as Sound: Emily Dickinson's Rhyme.* Athens: University of Georgia Press, 1990. Print.

Socarides, Alexandra. "For Emily, Wherever I May Find Her: On Paul Legault's Emily Dickinson." *Los Angeles Review of Books* (October 23, 2012).

Stamp, Jimmy. "A Brief History of Robot Birds." *Smithsonian Magazine*, May 22, 2013.

Stewart, Susan. *Poetry and the Fate of the Senses.* Chicago: University of Chicago Press, 2002.

Tiffany, Daniel. *Toy Medium: Materialism and Modern Lyric.* Berkeley: University of California Press, 2000.

Vendler, Helen. *Dickinson: Selected Poems and Commentaries.* Cambridge, Mass.: Harvard University Press, 2010.

Wolff, Cynthia Griffin. *Emily Dickinson.* New York: Perseus, 1988.

CHAPTER 7

The Starling's Whistle: Autophilology and the Order of Osip Mandel'shtam's Birds

Holt Vincent Meyer

but	as	oh
it's	a	if
clear	starling	only
my	eating	I
big-mouthed	after	could
friend	tastes	whistle
we	away	all
aren't	with	through
allowed	nutcake	life

Перо—кусочек птичьей плоти.
The pen is a little piece of bird's flesh.

—OSIP MANDEL'SHTAM, "Conversation about Dante"

Life and Text on a Perch

It is as scandalous as it is fitting for the topic of this article that the life of Osip Emilevich Mandel'shtam at its beginning and at its end resembled that of a bird having to perch on the next best branch in order to avoid plummeting into the depths of oblivion. Born at the beginning of the 1890s as a Jew on the western edge of the Russian empire,[1] Russia's greatest twentieth-century poet died a "public enemy" in Stalin's camps. On the eastern edge of the USSR during the mass-murderous "purges" of the late 1930s, Mandel'shtam almost constantly felt the fierce winds of mercilessly cruel history. Mandel'shtam was one of the foremost innovators in articulating a relation between Russian literature, Western European literature, and world literature at the beginning of the twentieth century. His poetry therefore presents an especially important challenge in measuring the global scale of world literary change against the changing European coordinates of traditional models of comparative literature. This essay considers how Mandel'shtam's figuration of birds tracks a relation between Western European poetic forms and Russian discourses shaped by a fear and rejection of such forms.

Russian culture was at this time enslaved by a bureaucracy designed in concentric circles of officials outdoing one another in mimicking Stalin, the absolute leader at its center. An order was being established that made the poet's life and work practically impossible. Beginning in the late 1920s, Mandel'shtam, whose cultural home was St. Petersburg (transformed into a very Russian and thus Eastern Leningrad), had extreme trouble keeping a roof over his own and his wife, Nadezhda's, head. The party leader Nikolai Bukharin, later tortured and murdered in Stalin's purges, had arranged a trip to Georgia and Armenia for the poet and his wife in 1930. This led to a renewal of poetic activity (he had ceased writing poetry in the mid-1920s, turning to prose and translation). It is from this period that Mandel'shtam's untitled starling poem dates.

The poem reads:

Куда как страшно нам с тобой, Товарищ большеротый мой!
Ох, как крошится наш табак, Щелкунчик, дружок, дурак!

А мог бы жизнь просвистать **скворцом**,
Заесть ореховым пирогом,

Да, видно, нельзя никак . . .

(Mandel'shtam, *Polnoe sobranie sočinenij*, 2:168–169)

Much we have to fear,
Big-mouth beside me!

Our tobacco turns into dust,
nut-cracker, friend, idiot![2]

I, on the other hand, could whistle all through life, as a **starling**,
And eat life's bad taste away with nut pie.

But it's clear: no permission, no way . . .

The poem's addressee ("big-mouth," "nut-cracker, friend, idiot!") is understood to be closely associated with the poet's wife. It is important to note in this context that the "I" of the poem's micronarrative does not appear lexically as "I" but rather as part of a "we" (together with the addressee) and as part of a verb construction in which the pronoun is suppressed ("a [ya] mog by prosvistat' [. . .], zaest'"/"[I] on the other hand, could whistle [. . .] eat"). One might say that the conjunction "a" (meaning "and on the other hand") takes the place of the almost homonymic pronoun "ya" (meaning "I"), replicating the very erasure of the sovereign subject that is the basic condition of the poem's lyrical self-presentation.[3]

The starling plays the role of producer of a certain type of "whistle." The whistling is not characterized in any way, except that it is somehow analogous or equivalent to the eating of sweet things to suppress life's bitterness. This is, in turn, something "we" might hypothetically perform but are prevented from doing by some undefined imperative. It is thus left open what it is about the starling's whistling that makes it equivalent to covering the bitter taste with the sweet. Two possibilities arise: Either the sound of the bird is analogous to the sweetness, or it is the act of whistling itself (like eating, as well as speaking, performed with the mouth). If it is the sound, then the analogy could be premised either on the characteristic sound of a starling's whistle or on the imitative quality of its song, since starlings mimic sounds from their acoustic environment.

I argue that the mimicry of the starling's song is the likeliest referent and that this is implied by the historical context of Stalinist Russia. I argue, further, that the analogy to the sweetness of nut cake is deeply ambivalent. In order to survive in Stalin's Russia, it was imperative to mimic—to display ultimate conformity. Mindless repetition was the core of that cult of personality that came into full force in the early 1930s, reaching its apogee in 1937 (the high point of the terror in the purges that ensued). For a poet, it was imperative to join the screeching hysteria of sycophants replacing the original Bolsheviks who were being purged from top to bottom of the party hierarchy. It was from this that Mandel'shtam fled during his journey to the south of the USSR, and it was in this context that he wrote his starling poem. This frame of reference helps explain why the poem is so elliptical. Addressed in familiar tones to someone personally close, it was nonetheless written with no certainty of its ever being published. Emblematic of the poet's impossible situation, the starling's whistle might be mindless repetition or meaningless sound. In its ambivalence, it might be both, mocking all the possible bird forms that had figured throughout Mandel'shtam's poetic work up to that point.

Flight, Becoming-Animal, Authority, and the "Conversation about Dante"

> Bird songs: the bird sings to mark its territory.
>
> DELEUZE AND GUATTARI, *A Thousand Plateaus*

There is a moment in Deleuze and Guattari's *A Thousand Plateaus* (in the chapter on "The Refrain") in which the bird is said to sing in order to "mark its territory" (312). Birdsong is related to the refrain itself, which is said to be "territorial," to be a "territorial assemblage." They go on to explain that

> territory is in fact an act that affects milieus and rhythms, that "territorializes" them. The territory is the product of a territorialization of milieus and rhythms. It amounts to the same thing to ask when milieus and rhythms become territorialized, and what the difference is between a nonterritorial animal and a territorial animal.

Through song, birds are generally deemed "territorial." Yet for Deleuze and Guattari birds are also most ambiguous with respect to the deterritorialization of "becoming-animal." Earlier, taking up Olivier Messiaen's idea that there are "musician birds and nonmusician birds," they ask whether the "refrain" is "necessarily territorial": "Is it not already used for very subtle deterritorializations, for selective lines of flight?" (301–302). This speaks directly to the ambivalence toward various species of birds in Mandel'shtam's texts, particularly those of the 1930s.

In Mandel'shtam's work of this period, one text stands out as the most complex and at the same time the most elaborated self-account of bird figuration: the "Conversation about Dante," which was written "during the spring-summer of 1933 in Koktebel, in the Crimea, and was not published in the Soviet Union until 1967" (*Complete Critical Prose and Letters*, 677). Like the starling poem, it was the product of a flight from the centers of Soviet power to the south of the empire. In this essay, it is poetry itself that is being negotiated (Dante's, but also Russian poetry of Mandel'shtam's time). Birds appear as a complex accompaniment to the argument. Most scholars agree that the essay, though ostensibly about the *Divina Commedia*, shows Mandel'shtam commenting on his own poetry. In other words, "Conversation about Dante" is part of the strong autophilological tendency of Mandel'shtam's texts. What I have elsewhere called the "autophilological"[4] refers to the tendency of literary texts to assume the responsibilities and the authority of philological treatment normally thought to be the domain of institutional (that is, academic) critical commentary. Mandel'shtam's "Conversation about Dante" is an ostensibly heterophilological text on the Italian early Renaissance, but to a great degree it is an autophilological text coming from within Mandel'shtam's own writing. This is germane for the topic at hand for several reasons. Autophilology negotiates and often displaces the authority of the institution in its jurisdiction over literature—something that in Stalinism was a natural oppositional desire of a writer but also one that assumes the role of official authority and thus affirms it at the same time. The play of the inside and the outside is, of course, also a question that can be addressed to all mimicry, particularly that of animals, since it is usually assumed that mocking animals only outwardly copy acoustics without getting inside the meaning of the sounds.

Mandel'shtam's Dante text is based on a deeply ambivalent working through of positions of authority. I thus argue against those readings of the essay that establish a personal identification between the author and Dante. Clare Cavanagh, for example, in *Osip Mandel'shtam and the Modernist Creation of Tradition*, sees Mandel'shtam's Dante as a "ringleader" in a "poetics of opposition" (212) and, in turn, associates this with Mandel'shtam's processing his Jewishness.[5] In "Keeping Time: Reading and Writing in 'Conversation about Dante,'" a salient study on the negotiation of authority, Jacob Emery[6] comments on the trope of the conductor's baton, a key image Mandel'shtam associates with Dante. Emery asks in general:

> How can we relate Mandel'shtam's depiction of angelic dictation to his documented awareness of writing's use, or misuse, as a tool of political control? And how can we approach poetry with the "conviction" and "trustfulness" he recommends, when the "hegemony of the baton" that culminates the reading process is identified not just with dictation but dictatorship? (508)

Emery continues, quoting from Mandel'shtam's essay:

> "Insufficient respect for the poetic material which can be grasped only through performance, only through the flight of the conductor's baton—this was the reason for the universal blindness to Dante, to the greatest master and manager of his material, to the greatest conductor of European art." The central metaphor of the conductor's baton (*dirizherskaia palochka*), to which generations of readers, like an unskilled orchestra, have failed to pay attention, occupies a nodal place in the essay's figurative structure, appearing at the climax of an ideal reading process in the sixth of its eleven sections. (495)

The use of the word "baton" itself is a call to rhythmic mimicry, something that Mandel'shtam applies to the rhythm of poetry itself. Emery's analysis underscores Mandel'shtam's claim that Dante invokes the total multimedia performance of poetry:

> If a finished poem is indeed "no more than a calligraphic product" that reaches its reader in a future far from the aegis of the poet, then under

> what authority is "the hegemony of the baton," which restores the text to the lived time of performance, generated? What power of compulsion or will maintains that hegemony, and what are the dangers in negotiating sight and sound, graphic image and dynamic performance? (497)

In answering these questions, Emery reads Mandel'shtam's text as

> an attempt to define the authority that regulates poetic performance in relation to other forces, sometimes overtly political ones, and to elaborate a theory of poetic notation adequate for the perpetuation of the text beyond the zone of the poet's personal influence. (497)

The issue of "force" is one of authority (and thus the call to mimic hegemonic order), and the perpetuation of the text is, among other things, a central task of traditional philology. Emery is thus in effect pointing to what I have been calling autophilology: the text itself assuming the functions and the authority of outside forces of order.

Mandel'shtam Counts Dante's Birds

Studies of Mandel'shtam's "Conversation about Dante" refer occasionally to birds in general (Emery in their analogy to angels and their relationship to God's authority), but none of them, indeed no one in Mandel'shtam scholarship, has looked closely at the appearance of types and species of birds in the essay, although the most salient passage is sometimes quoted for its commentary on Dante's "similes." Here is the passage in the original:

> Дантовские сравнения никогда не бывают описательны, то есть чисто изобразительны.
>
> Они всегда преследуют конкретную задачу дать внутренний образ структуры, или тяги. Возьмем обширнейшую группу "птичьих" сравнений—все эти тянущиеся караваны то журавлей, то грачей, то классические военные фаланги ласточек, то неспособное к латинскому строю анархически беспорядочное воронье,—эта группа развернутых сравнений всегда соответствует инстинкту паломничества, путешествия, колонизации, переселения.
>
> (*Polnoe sobranie sočinenij*, 2:168–169)

Interestingly and somewhat vexingly, the individual bird types are translated differently in the two English translations of this passage. They are indeed to a certain degree open for interpretation with a view to both the individual Russian and to the sources in Dante.

In translation by Clarence Brown and Robert Hughes we read:

> Dante's similes are never descriptive, that is, purely representational. They always pursue the concrete goal of giving the inner image of the structure or the force. Let us take the very large group of bird similes, all those long caravans now of cranes, now of crows, and now the classical military phalanxes of swallows, now the anarchically disorderly ravens, unsuited to Latin military formations, - this group of extended similes always corresponds to the instinct of pilgrimage, travel, colonization, migration. (*Selected Poems*, 11a8)

In the translation by Jane Grey Harris and Constance Link, the individual bird types (all appearing in swarms and formations) get a different treatment:

> Dante's similes are never descriptive, that is, purely representational. They always pursue the concrete task of presenting the inner form of the poem's structure or driving force. Let us take the very large group of bird similes, all of them extensive caravans now of cranes, now of grackles, now of swallows in classical military phalanxes, now the anarchically disorderly crows so unsuited to the Latin military formation . . . (*Complete Critical Prose and Letters*, 410)

Both translations use the term "anarchically disorderly" for a certain type of bird (formation), but the names of the birds shift. To be more exact, the "crow" changes places, giving way to the "grackles" of Jane Grey Harris and Constance Link. The accompanying table shows the shifts and differences.

Original	*zhuravli*	*grachi*	*lastochki*	*voron'e'*
Brown/Hughes	cranes	crows	swallows	ravens
Harris/Link	cranes	grackles	swallows	crows

There is a possible link between the "grach" and the "starling" through the "grackle," which, in addition, has a phonetic commonality with "grach." Although the word "starling" (*skvorets*) does not appear here, the "grackle," following one of its definitions, provides another term for an Asian mynah or starling with mainly black plumage. The Russian "grach" is, however, normally translated into English as "rook," and the Russian "voron" usually as "raven." It would seem, in any case, that "grach," following the order of the words for each bird formation, would be associated with the "anarchic disorder" linked explicitly to "voron'e" (swarms of ravens or crows, depending on the translation), while the cranes (*zhuravli*) are associated with the swallows (*lastochki*), who, in turn, form "classical military phalanxes" (both translations agree on this wording). The contrasting order and disorder of birds, in any event, seems more important than the particular ornithological classifications of the birds themselves.

The "similes" Mandel'shtam finds take on their own bird life. They cease to be simply illustrations of a particular type of approach to poetic imagery. They become poetic images. At least at first glance, the "caravans" of four different types of birds are broken down into two types: the orderly "Latin" ("Western") and the "anarchically disorderly" (and thus "Eastern"). It is no easy task to link these two categories and the four birds to the four categories "pilgrimage, travel, colonization, migration." Given the direct reference to the "classic military" and the "Latin military," it is possible to hear echoes of the "war machine" of Deleuze and Guattari's *A Thousand Plateaus* (especially, but not only, in the chapter on "nomadology" [351ff.]). The reference to the "Latin military" strongly suggests a form of ("Western") imperial configuration.

Here one can recall the figure of the cranes in the poem "Insomnia. Homer," from a time when Mandel'shtam's bird figures seem to follow along the more traditional lines of neoclassical tropes. The appearance of the "anarchic disorder" in the Dante essay, linked to the "ravens" (or "crows"), creates a tension, a chiastic ordering of meaning that might be linked to "travel" and "migration" (as opposed to the more ordered "pilgrimage" and "colonization"). It is conceivable that the "grachi," assigned neither to order (as is the case with the swallows) nor to disorder (as is the case with the ravens), can be associated with the figure of the starling found in the late starling poem discussed earlier. In that case, the "grach" (grackle or crow) would switch to the side of disorder.

Bringing all this to bear on the appearance of the "starling's whistle," I would again note that the unambiguously negative, "oppositional" association with the starling as a possible way of life (or of art) must be called into question. This in turn complicates the opposition order-disorder (suggesting a contrast between regular army and partisan formation), especially as concerns the political context of the poem in Stalinist Russia. Order attains an extreme ambivalence here, since the starling's mimicry tropes the conformity to an order of power in which the poem sees "no way" of participating. It seems that the starling represents an order that is really disorder, this then being a very special and complex instance of starling mimicry, stressing its undifferentiating nature, its mocking of other birds, animals, and even machine noises. The seemingly clean dichotomy between order and disorder breaks down in the starling's "whistle," which insists on the ambiguity of mimicry itself. Mimicry mimics the powers of order and disorder alike. Not simply oppositional, it is part of a complex game in which authority and order practically deconstruct themselves.

Mandel'shtam's Lyric Ornithology

Bird figures in general begin to appear early on, starting with one of Mandel'shtam's very first poems, and they remain a key factor in his very last poems. Here is one of his first known texts, dated 1908:

Звук осторожный и глухой
Плода, сорвавшегося с древа,
Среди немолчного напева
Глубокой тишины лесной.

(*Polnoe sobranie sočinenij*, 1:29)

The shy speechless sound
Of a fruit falling from its tree,
And around it the silent music
Of the forest, unbroken.

(*Selected Poems*, 3)[7]

A poem from October 1930, the same month as the one in which the starling text was written, begins with the lines:

Don't say a word to a soul.
Forget all you've seen,
Bird, old woman, cage,
and all the rest [. . .]

(*Selected Poems*, 57)

A text from the end of 1936 begins with this first of two stanzas:

Today is all beak and no feathers
and it's staying that way. Why?
And a gate by the sea gazes at me

Out of anchors and fogs [. . .]

(*Selected Poems*, 83)

A text from several months later, from March 1937, begins as follows:

How I wish I could fly
where no one could see me,
behind the ray of light
leaving no trace [. . .]

(*Selected Poems*, 96)

Another one from the same time begins:

I've lost my way in the sky—now where?
Let the one with the sky nearest to him answer.
It was easier for Dante's nine fallen
discuses to ring.

(*Selected Poems*, 94)

In the text of 1908, the "sound" amid the "constant" or "unbroken song" of the "woods' deep silence" clearly refers to birdsong, albeit in the form of its absence. Silence sings and thus becomes analogous to what usually sings in the forest: birds—appearing collectively absent. One might assume that the falling fruit is a trope for the Mandel'shtamian poetic word itself, the concrete simplicity characteristic of acmeism, the birdsong in this case being the flowery poeticism of the Russian symbolists (Briusov, Blok, Belyi) who were slowly losing their hegemony over Russian poetry after their dominance since 1890.

The texts of the 1930s evoke images of deficient birds, birds lost in the sky, or birds that have to be forgotten. Among the first birds to establish an ornithological paradigm are the cranes in formation in the poem "Insomnia. Homer," from the first collection, *The Stone*. Here the Homeric list of warships is rendered by the trope of cranes:

Бессонница. Гомер. Тугие паруса.
Я список кораблей прочел до середины:
Сей длинный выводок, сей поезд журавлиный,
Что над Элладою когда-то поднялся.

Как журавлиный клин в чужие рубежи—

(*Polnoe sobranie sočinenij*, 1:284)

Insomnia. Homer. Taut sails.
I've read to the middle of the list of ships:

The strung-out flock, the stream of cranes
that once rose above Hellas

Flight of cranes crossing strange borders [. . .]

I translate as "stream" (*poezd*) what Dmitrii Smirnov's renders as "caravan,"[8] suggesting an association with the cranes in Mandel'shtam's "Conversation about Dante." In this poem the cranes trope the ships listed in the Homer volume that the lyric subject is reading in his sleeplessness. The ships and thus the cranes, too, are figures of military conquest, well-ordered units combined for effective attack, "moved by love" (in the rivalry for Helen that is the occasion for the Trojan War).

The first entrance of the starling into Mandel'shtam's poetry is in the second edition of the same volume, *Tristia*, dated five years later (1920). The key stanza is as follows:

Где милая Троя? Где царский, где девичий дом?
Он будет разрушен, высокий Приамов скворешник.
И падают стрелы сухим деревянным дождем,
И стрелы другие растут на земле, как орешник.

(*Polnoe sobranie sočinenij*, 1:116)

Where is my dear Troy, where's the palace, the women's hall?
The tall starling-coop of King Priam is lying in shatters
And like a dry rain wooden arrows continue to fall
And more arrows just like a nutgrove arise in tatters.[9]

Here starlings appear addressed collectively as former inhabitants of a "coop," thus metonymically signified by a container from which the destruction of war frees them. The birds seem to be associated with the "wooden arrows" haphazardly falling in that same Trojan war, like rain. Then there is the blind swallow in the well-known poem of 1920: "I forgot the word I wanted to speak" ("Ia slovo pozabyl, chto ia khotel skazat"), which is sometimes given the title "Swallow." The swallow, appearing individually, is associated here with the forgotten word, the trope for the forgetting of the word being the "return to the hall of shadows."

Among individually appearing birds, the goldfinch deserves particular attention. In the mid-1930s, this bird emerges in full force in two of Mandel'shtam's poems which turned out to be among his last. In an earlier poem from 1922, little goldfinches are seen falling out of their nest. In the poems from 1936, the individual goldfinch is lovingly addressed in all of its eccentricity, and critical readings of these poems often take the goldfinch as a troping of the poetic self, a figure of ambivalent and perhaps self-mocking identification.

> Мой щегол, я голову закину,—
> Поглядим на мир вдвоем.
> Зимний день, колючий, как мякина,
> Так ли жестк в зрачке твоем? [. . .]
>
> (*Polnoe sobranie sočinenij*, 1:209)

> My goldfinch, I'll cock my head,
> together we'll look at the world:
> the winter day jagged as stubble,
> is it as rough to your eye as it is to mine? [. . .]
>
> (*Selected Poems*, 82)[10]

It has been noted in scholarship on these poems that the "I'll cock my head" refers to a gesture associated with Mandel'shtam's style of reading his own poetry for others (Surat, "Ėtjudy of Mandel'štame," 5). With these poems in mind, Joseph Brodsky in *Less Than One* writes:

> His became a poetry of high velocity and exposed nerves, with numerous leaps over the self-evident with somewhat abbreviated syntax. And yet in this way it became more a song than ever before, not a bard-like but a birdlike song, with its sharp, unpredictable turns and pitches, something like a goldfinch tremolo. (134)

Taking these two observations at face value, one comes to the conclusion that the goldfinch is a self-representation of poet and poetics: an animal (self-)philologist whose individual sovereignty is just as forcefully stated as it is eccentric.

Without being able to pursue this treatment of the goldfinch in more depth, I note that one could work through these two goldfinch appearances as figures of identification for which the individual starling—usually a collective bird (in its mass appearance and in its sonic mimicry of others)—serves as foil in the starling poem of 1930. In the dire time of the 1930s, these two birds—goldfinch and starling—represent contrasting figures for a troping of the poetic self: torn between mimicry on the one side and individual sovereignty on the other.

Revisiting "What could be more frightening for us"

Returning to Mandel'shtam's starling whistle, consider now a further translation by Dmitrii Smirnov: "O, we are so afraid, I and you, / My friend with the big beak! // Ah, our tobacco is crumbling, / O, Nutcracker, my dear, little fool. // One could whistle through life as a starling, / And eat a nut cake after it, // But we can't do it—it's clear . . ."[11] One particular feature of this translation is that it renders "bol'sherotyi," literally "big-mouthed," as "with the big beak," thus making the addressee into a bird as well. It is true that "bol'sherotyi" is a part of Russian animal names, but mainly of fish, not of birds, for example, in the Russian name of the largemouth bass (*bol'sherotyi okun'*) or the bigmouth buffalo (*bol'sherotyi buffalo*). For this reason, I believe the "beak" to be an overinterpretation. At the same time, the translation does call attention to the fact that the poem's addressee (and thus the lyrical subject as well) could be associated with an animal of some kind. The humans negotiated in the text would thus have some sort of zoological interface (where the border between the animal and the human is elided). This encourages the reader to look for paths of equivalence between the human and the animal and then also to ask what is the nature of the prescription with which the poem ends.

In saying that "we" could hypothetically be starlings (but then asserting "we" cannot), the poem opens the question of just who or what "we" are, zoologically or ornithologically speaking. But just where is the force that dictates "*no permission, no way* . . ." or "we can't do it"? And why is this "clear"? In answering this question, one has to account for the extremely emphatic doubling of negativity: not just "nel'zyá" ("it's not permitted") but rather "nel'zyá nikak" ("*no permission, no way*"). This leads to the next question: How literally is one to take the predicative adjective "vidno" (literally: "it is visible"; colloquially: "you see," "it's clear," "it's apparent")? What exactly is it that becomes "visible" or "clear," and why is it so clear? The text asserts this as if it were some kind of natural law.

At precisely this time Mandel'shtam was inspired by his acquaintance with the biologist Boris Sergeevich Kuzin in Yerevan, Armenia, where the poet had fled from the stifling capitals of Leningrad and Moscow. The most famous product of this acquaintance, which is said to have prompted Mandel'shtam to take up writing poetry again after a long break, was the poem "Lamarck." This poem marks the synthesis of natural science (particularly biology) and world poetry that formed the background of Mandel'shtam's later work. One might say that this embeds Mandel'shtam's humanist ornithology of the 1920s into a zoology entangled in the visions of a posthumanism that Stalinism makes omnipresent not only in the USSR in general but also in the poet's life (by presenting him and his wife with undignified living conditions).

The significance of this can be clarified and amplified by referring once again to the Deleuze-Guattari paradigm as concerns not only figures of flight but also, and above all, the central Deleuzian issue of "representation," which is seen to copy and reproduce the ever same, while other techniques of action and expression associated with the "rhizomatic" are depicted as a liberation from representation. In this sense, the mimicking posture of the starling is opposed to figures of eccentricity associated with the goldfinch. A chiasmus emerges in the relation between the mimicking starling and its eccentric opposite, on the one hand, and the Oedipal order and flight from it, on the other hand. The starling and its counterpart, the goldfinch, when linked to order or disorder, leave the respective other part of the binary opposition to the other bird. If the mimicking starling represents order, then the starling's eccentric nonmimicking other must by logic of the chiasmus represent anarchic disorder, and vice versa. The goal of the flight—be it an escape from order or an embracing of order—has the function of an unknown in this chiastic equation. The order of poetry performed by the verse itself sketches a field beyond articulation and denounces order itself as posthuman(ist), as an ensemble of babbling mindless beings swallowing down their lives with the help of sweet cake. So it is that positivity becomes negativity and negativity becomes positivity, for the alternative to the mimicking starling is invisible and unsayable but necessary. This recalls the ambivalence of order and disorder addressed in connection with Mandel'shtam's Dante essay.

Here is the import of the doubly emphatic "nel'zyá" (one cannot, one must not), which strives for—or flees to—a visibility beyond what is actually visible (that is, allowed to be visible and sayable by the discursive forces shaping the public sphere in the USSR in 1930). If one assumes that the starling is a figure of mimicry and that the prescription is connected to the categories just outlined, then one can conclude that the imperative not to copy the copier associates the mimicking of the starling with both types of order and thus, in all its ambivalence, attempts to open up a third space that is neither or both. It is into an animal that does this that the "we" are supposed to evolve.

There is also a parallel here to the Lamarckian "soft evolution," which allows for nonbiological factors in evolution (that is, the inscribing of history into the discourse of animals) and was taken up by ideological Stalinist pseudoscientists like Lysenko. Evolution can also be read as adaption and thus mimicry over a long time period. In our context, this takes on a whole new meaning: a concept of evolution that has nature coming unhinged and getting subordinated to the social. Mandel'shtam's "nel'zyá"—"it is forbidden"—with respect to becoming-starling produces a parallel discourse to that of evolution—a poetic and poetological metamorphosizing that rewrites the history of literature as a becoming-bird.

The starling as a copy-bird represents in the poem (*nel'zyá!*) a flight path (in the literal sense of an actual bird's flight pattern) forbidden to the "we" comprising the lyric subject and his interlocutor. At the same time, the interdiction itself represents a flight (in the Deleuzian sense of a flight from the oppressive Oedipal order), a flight from mimicking, which is possibly associated with the emerging Stalinist system. This flight, also figured in the bird formations from the Dante essay, is marked both as order and as disorder, both as army and partisan, and seems to open up an unnamed third space that can only be designated negatively. Mandel'shtam's starling whistle can be related to poetry as a technology, or *techne*, which crosses over into political, biological, and other discourses to resist, and at the same time participate in, the posthuman or the nonhuman.

NOTES

1. Mandel'shtam was born to Jewish parents in Warsaw and as a Jew could only study in the capital, St. Petersburg (or anywhere outside the "pale of settlement"), by converting to Christianity (to the Lutherans in Finland), that is, by bending to the force of the authorities. Clarence Brown (*Mandelstam*, 46), referring to a comment by Mandel'shtam's wife, Nadezhda, notes that he was "baptized a Christian at some Lutheran church in Finland, a step that removed him from the Jewish quota but one to which he later attached no significance at all."

2. These first four lines are taken from the translation by Clarence Brown and W. S. Merwin (*Selected Poems*, 57). The following lines are my own translation. Brown and Merwin translate as follows: "And I could have whistled through life like a **starling**, / eating nut pies . . . / but clearly there's no chance of that" (emphasis added).

3. In 1928, Mandel'shtam famously answered an official Soviet questionnaire, whose first question addressed the effect of the revolution on the poet's work, with the extremely ambiguous words: "It was inevitable that the October Revolution should influence my work since it took from me my 'biography,' the sense of personal significance. I am grateful to it for having put an end once and for all to my spiritual prosperity and living on cultural dividends. . . . I feel myself indebted to the Revolution, but I bring it gifts of which, for the time being, it has no need." In *Contemporary Poetry: A*

Retrospective from the Quarterly Review of Literature, ed. Theodore Russell Weiss and René Weiss, trans. W. S. Merwin and Clarence Brown (Princeton, N.J.: Princeton University Press, 2015), 514.

4. I have tried to work out basic features of "autophilology" in the articles of mine listed in the Works Cited section. In "Čtení Haška s Deleuzem," as in the present essay, I attempt to link these issues to the thinking of Deleuze and Guattari.

5. See also Suchoff, *Kafka's Jewish Languages*. For another account of Mandel'shtam's Judaism, one connected to Benjaminian models of translation, see Meyer, "Chaos iudejskij."

6. Recent scholarship shows that Mandel'shtam develops an oppositional position in the Dante text and the writing of the 1930s. Jacob Emery's study is one of three studies on Mandel'shtam's Dante text (the others being by Julia Vaingurt and Alexander Spektor) that appeared in the *Slavic Review* (Fall 2014). Spektor's working through of Mandel'shtam's strong interest in the natural sciences in the 1930s and Vaingurt's treatment of the cultural location and the cultural modeling of the "Conversation about Dante" are also of interest for my topic. Vaingurt comments on one of the few places where Mandel'shtam refers to Dante's *Paradiso* (as opposed to *Inferno*, from which most of the Dante material is quoted and discussed). Mandel'shtam picks up on the passage in Canto 18 where birds (*augelli*) form the letters "D," "I," and "L" (the first letters of "diligite justitiam") and notes that the grammar of Russian, like that of Old Italian, is like that flock of birds, which in turn is similar to the "Florentine mob which changes the laws like gloves." Instead of syntax, Mandel'shtam continues, there is a magnetic pull and a yearning for Florence. What Vaingurt does not explicitly mention is the birds' visual mimicry of writing, which adds another aspect to the general issue.

7. On the Internet one finds this alternative translation attributed to Andrey Kneller: "The careful muffled sound of fruit / That plummets, broken from a tree, / Amid the constant melody / Of the deep silence of the wood . . ." https://sites.google.com/site/poetryandtranslations/osip-mandelstam/the-careful-muffled-sound-of-fruit.

8. "Insomnia. Homer. The rows of stretched sails. / I've read the catalogue of ships just to the middle: / That endless caravan, that lengthy stream of cranes, / Which long ago rose up above the land oh Hellas. // It's like a wedge of cranes towards the distant shores . . ." https://en.wikisource.org/wiki/Insomnia.Homer.Therowsofstretchedsails.

9. Translation by Ilya Shambat: http://www.lib.ru/POEZIQ/MANDELSHTAM/tristiaengl.txtAscii.txt.

10. Alternative translation: "My goldfinch, I'll throw back my head, let's look at the world together: the winter's day is prickly like chaff, does it seem

as harsh to your eyes?" Osip Mandelstam, *The Voronezh Notebooks*, trans. Richard McKane and Elizabeth McKane (Newcastle: Bloodaxe, 1996), qtd. at http://blindflaneur.com/2007/10/09/osip-mandelstam-the-goldfinch/.

11. https://wikilivres.ca/wiki/O,_we_are_so_afraid,_I_and_you.

WORKS CITED

Brodsky, Joseph. *Less Than One*. New York: Penguin, 1987.

Brown, Clarence. *Mandelstam*. Cambridge: Cambridge University Press, 1978.

Cavanagh, Clare. *Osip Mandel'shtam and the Modernist Creation of Tradition*. Princeton, N.J.: Princeton University Press, 1995.

Deleuze, Gilles, and Félix Guattari. *Kafka: Toward a Minor Literature*. Minneapolis: University of Minnesota Press, 1986.

———. *A Thousand Plateaus*. Minneapolis: University of Minnesota Press, 1987.

Emery, Jacob. "Keeping Time: Reading and Writing in 'Conversation about Dante.'" *Slavic Review* 73, no. 3 (Fall 2014): 494–513.

Eshelman, Raoul. *Nikolaj Gumilev and Neoclassical Modernism: The Metaphysics of Style*. Frankfurt: Peter Lang, 1993.

Friedmann, H. *The Symbolic Goldfinch: Its History and Significance in European Devotional Art*. Washington, D.C.: Pantheon, 1946.

Hitzke, Diana: *Nomadisches Schreiben*. Frankfurt: Peter Lang, 2015.

Mandel'shtam, Osip. *The Complete Critical Prose and Letters*. Ed. Jane Gray Harris. Ann Arbor: Ardis, 1979.

———. *Polnoe sobranie sočinenij*. 3 vol., Moscow: Progress-Plejada, 2009–2014.

———*The Selected Poems*. Trans. Clarence Brown and W. S. Merwin. New York: Macmillan, 2004. (First published in 1973.)

Meyer, Holt. "Der Akmeismus Mandel'štams als eine modernistische Neoklassik." *Wiener slawistischer Almanach* 107–147 (1991).

———. "Chaos iudejskij: Mandel'štams Übersetzungskonzeption aus kulturologischer Sicht." In *Juden und Judentum in Literatur und Film des slavischen Sprachraumes. Die geniale Epoche*, by P. Kosta, N. Drubek-Meyer, and H. Meyer, 193–226. Wiesbaden: Harrassowitz, 1999.

———. "Čtení Haška s Deleuzem: Devenir-animal a česká autofilologická deteritorializace." In *Sborník příspěvků z IV. kongresu světové literárněvědné bohemistiky*, ed. L. Jungmannová, 341–350. Prague: Ústav pro českou literaturu, 2010.

———. "Dlouhé loučení Milana Kundery s Julkem Fučíkem a odpovědnost žánru. První verze Umění románu a druhá verze Posledního máje." In *Slovo a smysl*, 23:36–65. Prague, 2015.

———. "Durch den Regenbogen schweben: Klammer-Setzungen in Nabokovs Dar als autophilologisch sprengende Rahmen." In *Den Rahmen sprengen: Anmerkungspraktiken in Literatur, Kunst und Film*, ed. B. Metz and S. Zubarik, 63–89. Berlin: Kadmos, 2012.

———. "Eine Episode aus Mandel'štams 'Stummen Jahren': Die Max-Barthel- Übersetzungen." *Welt der Slaven* 36, nos. 1–2 (1991): 72–98.

———. "The Inside of the Outside of the Untranslatable (Bracketed Pushkin) in Nabokov's Gift." In *Les intraduisibles / Unübersetzbarkeiten*, ed. J. Dünne et al., 73–89. Paris: Éditions des archivs contemporaines, 2013.

———. "(Klammer)Übertragung und (Theorie)Widerstand im zweiten Kapitel von Nabokovs Gabe als autophilologische Techniken im Rahmen von Freud- Vater- Behandlungen." In *Rhetorik der Übertragung*, ed. D. M. Nielaba et al., 121–138. Würzburg: Königshausen und Neumann, 2013.

———. "Я: вот видите—Unreadabilities and/as Autophilology in (Prigov's) Letter Work: Onegin as an Alphabet and the Azbuki Hg. Brigitte Obermayr." In *Wiener Slawistischer Almanach*, 81:72–161, 2013.

Spektor, Alexander. "The Science of Poetry: Poetic Process as Evolution in Mandel'shtam's 'Conversation about Dante.'" *Slavic Review* 73, no. 3 (Fall 2014): 471–493.

Suchoff, David. *Kafka's Jewish Languages: The Hidden Openness of Tradition*. Philadelphia: University of Pennsylvania Press, 2012.

Surat, Irina. "Ėtjudy of Mandel'štame." *Znamja* (2007): 5.

Vaingurt, Julia. "Mastery and Method in Poetry: Osip Mandel'shtam's 'Conversation about Dante.'" *Slavic Review* 73, no. 3 (Fall 2014): 457–470.

CHAPTER 8

Colonial and Postcolonial Birds of Game, Games of Bird

Fawzia Mustafa

from
the
air

planes
of
serengeti

wildlife
concentration
camps

what
I
see

when
the
birds

of
theory
shit

the
seagulls
sound

like
newborn
dead

off
robben
Island

Yvette Christiansë's novel *Unconfessed* (2006) is set in the Cape Colony and opens around 1826, just after the new British administration instituted "amelioration" measures into slave governance in preparation for eventual emancipation. Bernhard and Michael Grzimek's documentary film *Serengeti Shall Not Die* (1959) is set within a contemporaneous Tanganyika, as the colonial government was establishing the boundaries for the famed Serengeti National Park. Along with the transfer of power between Dutch and British control of slavery, the novel's historical backdrop is also the moment of the transformation of certain animal species into "game," while the film enacts the transformation, more than a century later, of (neo)colonial land policy into "game parks" driven by a curiously antihuman logic of "conservation." Juxtaposing the works allows for the genealogy of these transfers of power to emerge as central and crucial strategies of colonial control designed to withstand and outlive formal colonization. At the same time, their contingency also exposes an often overlooked relation between colonial oppression and animal rights and, today, a tension between posthumanism and postcolonial theory. While the two works have nothing obvious in common, I hope to show that they are indeed connected through the trajectory of archival figures, contingent colonial capitalisms, and legislations that transfer value from (some) humans to animals, which is revealed in part through a series of figurative plays between games of birds and birds of game.

Yvette Christiansë's novel *Unconfessed* undertakes a fictional rendering of a slave's subjectivity and consciousness at the time of the transfer of power between Dutch commercial and British colonial administrations in the Cape Colony. Drawn from rather than inspired by snatches in the archives, where a record of Sila Van de Kaap's trial for the murder of her son Baro is housed, the fictional protagonist, Sila, and the figure in the archive, Sila, are caught between the Roman-Dutch and emerging British laws regarding the status of slaves.[1] This situation allows the novel to function politically as an intervention into the historical record, interrogating the legal premises of the colonial archive. This intervention cleverly mirrors the slave strategy outlined by John Edwin Mason (borrowing from James Scott), whereby resistance to slavery is asserted within the terms of slavery itself, within "the 'moral economy' of the lash." Mason's coinage is specific to the corporal punishment of slaves, which he explains as "a moral code which acknowledged the legitimacy of physical punishment but set rules governing its distribution, [and] the circumstances under which it could and could not be justifiably applied" (Mason, "Hendrick Albertus," 428). Sila's moral economy is defined by the legal framework first of Roman Dutch laws, then of emergent British law. In both cases, Sila's resistance is internal to the law of slavery rather than an assault upon the institution itself. Within the novel and in the partial records, under the prior Roman-Dutch law and in her owner's will, Sila had gained her freedom but nevertheless falls victim to a common practice of deceit, and she remains enslaved by her owner's descendants. She spends long years being shunted between enslavement and incarceration as her case for murder is variously adjudicated. Much of the novel is staged through her incarceration on Robben Island and within the urban Cape's prison, making the island and the town complex settings for major parts of the novel. The other locations within the novel's flashbacks are the slave-owning Dutch farms, also the setting and occasion for the predominant Afrikaner literary genre of the *plaasroman*, the "farm novel."

Mock Mock

Another politics also underwrites the novel's narrative choices and encompasses the expressive area I want to concentrate on in this essay. In the novel, as Sila recalls her life, two contingent narratives unfold: one in the third person, within the novel's narrative present, and the other, a first-person internal monologue often addressed to the dead son, Baro, and other absent characters, which offers both information about the past as well as commentary on the shifting present. So while the novel gives fictional space to Sila, its way of doing so is far more complex than simply an imaginative reconstruction of her life might suggest. As Rosalind Morris has pointed out, "We must not take the fiction as a 'speaking for'; it is rather a speaking in the absence of any fixed place from which utterance could claim to be grounded in identity." As Morris goes on to argue, "the Sila of *Unconfessed* is an entirely unreliable narrator whose broken, recursive, and often contradictory discourse evades the desire for unity at every turn" ("In the Name of Trauma," 405–406). In her own reading of the court documents from which she draws, Christiansë points out that Sila's archival visibility is only as a "shadow figure" whose "stubborn presence [nevertheless articulates] a desire for speech resulting from the inability to be heard fully from within slavery's discourse" ("'Heartsore,'" 1). Christiansë continues to suggest that Sila's position within the archive is akin if not identical to the one famously described in Gayatri Chakravorty Spivak's "Can the Subaltern Speak?" Christiansë goes on to explain, "Sila [in the archive] is structurally muted in that, although we have words from her, the state never granted her full subjectivity, and her utterances remained, for them, utterly illegible" (1). At the level of the novel's plot, this remains the case: Sila's "interrogations" only operate in one direction, and the series of her incarcerations follow the course of a judicial system cementing only its own authority. As Leila Nabizadeh has pointed out:

> The pages of the novel's first chapter stand out from those of the rest of book because it is only in the first chapter that Christiansë uses quotation marks to indicate that a character is speaking and employs a third person narrator. The remainder of the novel is figuratively silenced because there is no indication that anyone is speaking and is told from the first person perspective of Sila. These characteristics of the dominant portion of the novel suggest that the transition in perspective is from a free, non-colonized observer to that of the subaltern. (2)

Christiansë's essay offers the following strategy for "reading" the archive:

> One must learn how to listen to echoes of subjects for whom one might not have an adequate language; one must also learn how to discern what they might have been trying to say within the statements

> attributed to them (but that could very well represent the redactions of colonial officials—notaries, court reporters). In addition, one must prepare to hear and interpret any echo of the unsaid as something that could be nothing more than a trace. (4)

This unspoken, unspeaking, and unconfessed "silence" within the novel is not a barrier to the ability to convey information and articulate complexity. Following Christiansë's lead, my reading is of two embedded critical devices, both of which entail the figure of birds, one without "voice" but with "sound" and the other without "rationality" but with "reason," which attempts to make this form of inarticulation and its critique clear.

The use of sound is both ambient and foregrounded, and the critique it offers mocks the elaborate and rigid patriarchal and paternalistic control that characterizes all the social formations underwriting this slave society. Toward the beginning of her sojourn on Robben Island, as Sila converses in her mind with Baro, one of her self-appointed tasks is to school him on the routines of incarceration on Robben Island and on the island world itself. Windswept in the extreme, the island sounds have already begun to emerge as a ubiquitous and ambient background roar. She says, "small as this island is, you may be pulled about by the wind or the sound of the sea—oh, and sometimes, it is not the sea that you hear at all, but the wind sounding like the sea. So, be awake. You will hear the difference when you go near the *fynbos*, you will hear wind and think it is the sea. But, remember, the sound is thinner" (*Unconfessed*, 44). Sila's ability to distinguish between these two elements signals first her ability actually to discern between the sounds of the wind when mediated by the heath (*fynbos*) rather than the sea and then her recognition of what it is that she discerns: sounds both signifying and mystifying their own *difference*. The mediation of *fynbos* soundings, furthermore, becomes metonymic of the seagulls' cry and the role it plays in both clarity and confusion, of place, of location, of situation, and, of course, of the sliding scale of sanity, not to mention humanity.

Sila offers Baro the above instructions within a specific context mediated by the figure of birds. The chapter opens with, "I woke up. *What?* It was the gulls. Gulls can cry like newborn babies. And newborn babies can sound like little creatures sent to us so that we may teach them how to become like us" (180). Sila has recently given birth to her eighth child, Debora, who, like an unnamed boy child earlier, will not survive infancy. But Debora is also among the four children born to Sila while she has been enslaved and incarcerated, a simple but grim fact that marks the scale of slavery's informal economy of human reproduction. The latter part of Sila's observation, accordingly, has less to do with child development and more with the exigencies of slavery's caste system. That three of her eight children die, albeit one by her own hand, also suggests a careful statistical rendering of both the infant-mortality rates of the era and its rates of abortion and documented and undocumented infanticide. The bay in which a contemporary of Sila's, Hester, tried to drown herself and her children was known for the baby carcasses and fetuses that washed up on its shores. Hester's case, like Sila's, exists in the archive, in histories, and now, through this fiction, in *Unconfessed*, too.

The gull, meanwhile, cries like a newborn baby, enough at least to startle momentarily the newly birthed mother into thinking it her child in distress. Whether coincidence or resemblance of sound in terms of pitch and duration, the gull's cry performs a remarkable effect in the story. Sila's identification of the cry and the seagulls' huge population on the island *silently* releases another level of both countersurveillance and haunting within the novel: gull cries as *heard* metonym for loud, lost, dying, and dead babies serve as an insistent locator of a proper count of the colony's dead babies and the open secrets they signify.[2]

It is not surprising, then, that Sila chooses the seagull as the vehicle for two of her songs, embodying the bird's ability to remind, or to not forget, not only through its cry but as emphatically with its beak:

> *Seagulls pick today at the water's face, and the water's face dips away and will not be still and the fish leap and dive behind the seagulls' backs. . . . I have a mind to go where the fish go in that deep down there where seagulls are not welcome.* (213–214)

The seagulls' action, a kind of faceoff with the ocean, enables the novel's figurative play to employ the beak as more than the body part from which the gulls' cries are expressed; rather, the beak both fishes and probes, reading and writing the surface with glimpses of what's below the surface, but silently, even revealing a vulnerability to easy evasion and bypass. The song is interrupted by a direct address to Baro, where Sila recognizes the analogy she has just made: "Today I am pleased to put my own words to long use and my lips are pleased to push out and hold the door open for things that come from deep in the cave of my heart." She then sings the second song, "*What if it is not that the seagulls pick at the water, but that they pick today to undo the world? And why would they pick-pick, pick-pick today?*" (213–214), which is that disclosure or that discovery of access, that revelation of what the surface hides. In Christiansë's essay "'Heartsore,'" she explains the key word in the archive associated with the historical Sila as the Dutch word *hartzeer* or *hertzeer*, translated severally as "heartache" and "heart-sore." In one iteration, the word "has the status of a verbatim citation" but in another a purported record of Sila's own testimony admitting to the crime of killing Baro. The essay is meticulous in explaining the prevailing legal discourse's limitations and the overwhelming inability of the archive to settle its own record of what was meant, let alone what was heard, said, or ascribed: As Christiansë notes, "what remains is that which was and still is in excess of the law and, indeed, in excess of her," the "translitera-tion of the Dutch 'hartzeer'" (18–21). Resistant to even the idea of recon-struction, Christiansë unravels just this impasse in the novel, by maintaining the ambiguity as well as "excess" of the nonspeaking nonsubject. Hence, the secret of this status is both maintained and unfolded through its task, this "*pick-pick*." The novel performs this "*pick-pick*" by bringing together a series of contingent, even accidental historical players and events that help ex-pose the extraordinary symbolic and actual social landscape that Sila's eye and ear demarcate so accurately, undoing, as she claims, "*the world*."

By this stage, the novel has already performed another surreptitious sur-veillance in its careful study of the island's social formation. Just prior to the story of Hester that Sila tells Baro, pointing out the different modes of execution for women and men and offering an apostrophe to her, Sila makes a series of casual references about the island's other visitors, whites from the mainland who arrive for commercial and recreational purposes. The coordinates of a specific political economy begin to come together: Along with the sexual exploitation of the female prisoners by both wardens and male prisoners, a generalized prison labor itself is utilized to provide stone for the mainland, along with commercial fishing and recreational hunting (for rabbit) made available to the public. The sharp contrast of the island's division between incarceration and recreation particularly between black and white is unequivocally rooted in the emergence of a new economy of human-to-human and human-to–nonhuman animal relations. Accord-ingly, the development of animal capital alongside the human is surrepti-tiously explored. R. B. Fisher, in his letter informing Wilberforce of the impunity of infanticide in the colony, which set off the controversy about infanticide for the new British administration especially under the gover-norship of Lord Charles Somerset, makes an observation that draws on a well-worn analogy between the treatment of slaves and animals. He notes that slaves "therefore exist in a much less protected state than the hares and pheasants, and even the rabbits of this country" (in Theal, *Records of the Cape Colony*, 11:179). What is apposite here is the choice of animals:

Rabbits had been introduced to Robben Island in the mid–seventeenth century to help provide provisions for maritime traffic, but game birds such as "Pheasant, Partridge, Korhaan, Wild Peacock (Paauw,) Ostrich," and "Buck (comprising the Whole Antelope Species,) Hare, or Zebra" were among the very first species to fall under the first game laws passed anywhere in the British Empire. The conferral of the status "game" on some birds (and mammals) establishes the moral economy that allows for the emergence of a more fully capitalist moral economy of "big-game" politics. It is this moral economy, this new hierarchy, that the novel's subtle juxtaposition of incarceration and recreation recognizes. Somerset's Proclamation of 1822 laid the groundwork for all subsequent legislation over wildlife, hunting and blood sports, game parks, and conservation of animal species from this time forward.[3] I will return to this.

Who's There?

Sila offers occasional reflections on the people who come to Robben Island from the mainland to hunt, picnic, and fish. The first is when Sila learns that her death sentence has been commuted but that she must nevertheless serve a lengthy sentence on Robben Island for murder. Breaking the news to a fellow inmate who is also to be transferred from the prison in Cape Town to the island, Sila muses, "Robben Island. Nothing good was ever said about it except by white people who went over to picnic and catch fish" (18). Later, an equally nagging reference takes place on the island early in Sila's not quite apostrophe with the dead Baro. Her schooling him about the island includes one of Sila's many warnings:

> If you hear guns, it might be someone shooting rabbits. Visitors come to hunt rabbits. Or they come to catch fish. Some days, you will get used to this, you hear the laughter of visitors resting on the rocks after a good hunt. The men light pipes. The women sit under parasols. (46)

This is immediately followed by one of Sila's mental reflexes of sharp rationality, "*What kind of people are these?*" (46). These reflexes, shocks of recognition, are the narrative's way of measuring the degrees of operational dehumanization underway, and this in turn allows for the novel's parallel narrative of interiority to develop. Thus, when Sila recounts the occasion of Warden Pedder's visitors, who arrive to explore the commercial potential of both fishing and, conceivably, the development of a recreational site, the subsequent exchange, interspersed with Sila's asides, redounds into nothing less than a horrific sublime:

Dear Mr. Bowler, Mr. Pedder, if it were not for the unfortunate use to which this Island has been put it would surely be the most glorious fairground for the town.

Oh, yes, Mariah has it exactly. Mr. Pedder, do you not agree that the town is in need of a place where ladies and gentlemen could have the best balls and parties. It is so dull otherwise.

Dull? Ladies, ladies! What of the freedoms we all enjoy? In the midst of the wildflowers? Consider those that give us such brilliance and beauty in brief blooms that last but a day, though full and generous. Consider the nights we enjoy. So calm.

And so cold, Mr. Big Nose.

So clear.

So that some can see the mainland stretch into the length of their longings,

Mr. Bald Head Big Nose.

So serene. A serenity that, surely, puts one in mind of god's paradise.

And let us not forget that snakes go about at night, Mr. Flabby Crotch Bald Head Big Nose. And men.

Stars trembling . . .

The men tremble as their toes dig in and push, push.

. . . shivering

You can say that again!

The mainland floats into the dark like my home as we are loaded into the ship. We are taken out at night in long boats and we are pulled up in a big net and that net puts us in a night darker than the demon's heart.

(74)

Here, Sila's interjectory asides continue to operate as a historicizing element, even as the exchange she reconstructs knowingly taps and segregates the rhetoric of high romanticism (freedom, wildflowers, paradise) from the tableau of rape to the scene of transport after capture. The novel is careful to establish a contestation for every act of colonialist sublimation, pitting reason, logic, and fact against rationalization and ideology. All is encapsulated in her final riposte, where the repetition of "net" suggests the discursive binds that function as effectively as any physical incarceration: "*Do not come here talking nonsense, Mister*" (74).

Sila's uncanny insight is stripped of the distorting overlays of (colonial) fantasy, socialization, acculturation; indeed deranging its very logic. Nevertheless, or perhaps because of this, just after the gull's cry, Sila envisions her hair growing rapidly and massively, knocking on doors, entangling and eventually taking over the whole island. This startling scenario unleashes the complex imagery of discursive entanglement as well as, brilliantly, a figurative site for *nesting*. She then enjoins the future perfect, the actual grammar of which anticipates the tourists whom, surely, she and Baro will haunt:

> A big bushy forest filled with things that laugh and sing inside. People will come in their boats. The ladies of Cape Town will give up visiting each other and come out in boats that trail ribbons. . . . It will be as if they do not know what makes them say my name. And yours, Baro.
>
> They will say it is the wine and they will cover their mouths with their hands and apologize while they giggle. And my hair will send a bird or two out. They will cover their heads quickly and duck under their parasols because the birds that fly out of my hair will be known for the size and accuracy of their shit.
>
> *Dit is what ek sê!*
>
> (181)

Unlike the (literary) gulls and their cries, these birds are birds of theory, not the Cartesian "theoretical animal" but a non- or antianthropomorphic theory whose effect is active, not passive. Rather than being seen, they see, or, if they are "seen," they are only "seen" as bird shit.[4] The figurative play of these birds highlights a productive relay between a discourse of colonial nonsense and a discourse of postcolonial shit. Nonsense and shit (expressively synonymous, as in bullshit) remain actively and multiply signified as terms within Reason. Colonial nonsense is non-sense, as in Homi Bhabha: "The articulation of nonsense is the recognition of an anxious contradictory place between the human and the not-human, between sense and non-sense" ("Articulating the Archaic," 125). Bhabha's reading of the resistance, refusal, or inability to translate cultural differences locates colonialism's (uncanny) blind spots, the colonial signifier's suspension in the "in-between" of "neither one nor the other": between non-sense and nonsense (126–127). More simply colonial nonsense emerges from the stubborn resilience of colonially discursive regimes that prevail within the locked parameters of their political and economic control, allowing the fantastical and bizarre to stand in as the real: such as mute subalterns or a hierarchy of the "human"—that is, "race" wherein categories of partial humanity are admitted and sustained. Thus, Sila's interjections, rejoinders, in the long quotation above can be read as the inarticulation that almost displaces the real in the skewed and skewering symbolic order of a slave.

Postcolonial shit, as a response produced in the figurative play of the birds in Christiansë's novel, though structurally more oppositional in its always already deconstructive inception, also operates as a slippery category in its own history of alignments and displacements. On the one hand, it is the sword that cuts through the Gordian knot, defiant, reactive and counterhegemonic (such as Sila's words above); on the other, it is severally available to any number of articulations: anthropological (colonial "hygiene," "defilement," "contagion"), Freudian (wealth and greatness), capitalist (guano), and post-Freudian ("theory" itself, perhaps). Two possibilities stand out. Within the annals of slavery, Saidiya Hartman's stinging chapter "So Many Dungeons" in *Lose Your Mother* reminds us that the accumulated sedimentation lining the ruins of the holding dungeons for slaves awaiting the transatlantic middle passage was and remains "[Human] Waste [that] is the remnant of all the lives that are outside history" (115). Properly historicized, however, the equation that reveals itself from a reading that understands such Waste as postcolonial is: waste = shit = excess = surplus = meaning.[5] This constitutes a release, in other words, from "an anxious contradictory place," such as an archive. At the same time, the *figure* of postcolonial shit in *Unconfessed* is bird shit, not human, and therefore between signifier and signified it, too, plays with "an anxious contradictory place between the human and the not-human," inasmuch as it targets the human with the question of what *is* human if some are now, legally, considered *not*. The second is what Eduardo Cadava brilliantly teaches us: The melding and sedimentation of bird droppings and human and other animal remains, excremental and otherwise, is quite properly as well as figuratively the "guano of history," that is, the conditions of capitalist possibility wrought by violence and overwhelming oppression that, with its bird games, *Unconfessed* so uncannily discloses.[6]

Mocking!

By employing an animal turn of sorts, then, I want to pursue further a series of connections between *Unconfessed*'s exposure of wastes and their subsequent meanings.[7] Returning to the archive, we see some interesting connections: The figure of Somerset within the archive is the recipient of Lord Bathurst's inquiry about infanticide in the Colony after his receipt of Fisher's letter to Wilberforce. Denying any unusual statistics of infanticide, Somerset maintains, after an inquiry, his defensive mode in an effort to *forestall* in the Cape the more comprehensive postabolition slave reforms being formulated in Britain and about to be implemented in the Caribbean. As the presiding governor, Somerset, an unapologetic High Tory, issued the Proclamation of 1823. No provisions for enforcement were made, even while appearing to "ameliorate" slave conditions, which included banning the sale of slave children, offering some standing in courts of law, and allowing marriage and the right to own property; however, *every* item in the proclamation was qualified by the requirement of baptism in the Christian faith, making it prohibitively exclusionary. Somerset had taken office as governor in 1814 and was himself ambivalent about abolition and emancipation (despite his mother being "a champion of anti-slavery and a friend of the leading British emancipationist William Wilberforce"), though he remained answerable to enough members of parliament who were not (qtd. in Weaver, "Litigating for Freedom," 47). His was an embattled tenure and, in addition to the Proclamation of 1823, he was also known for being partially responsible for destabilizing the fragile frontier between Afrikaner farmers and the Xhosa people; for his autocratic style of rule, which fueled the changes instituted by the 1822 Commission of Inquiry over the governance of Britain's post-Napoleonic acquisitions; and, perhaps his most well-known act in the Colony, his attempts to censor the infant press of Thomas Pringle and John Fairbairn. Somerset's less-than-illustrious sojourns as governor, the first of which was until 1819 and the second from 1821 to 1826 (just prior to the passage of Ordinance 19), is also marked by his institution of the first game laws in 1822.

The proximity of the two proclamations (1822 and 1823) is less interesting than the disposition of the person who formulated them. Somerset's ambivalence over emancipation accounts for the false start that reform had in the Colony, while his enthusiasm for (hunting) animals helps explain the start of what would soon become a major enterprise on the continent. It so happens that his father was the Duke of Beaufort, "inventor of the modern version of the 'sport' of fox-hunting and whose country seat at Badminton boasted 116 rooms and innumerable multitudes of horses, dogs and servants" (Peires, "The British and the Cape," 473). While introducing the sport to the Cape (substituting the jackal for the fox), he became (more) concerned about the rampant slaughter of game by unregulated settler hunting practices. According to one source, in 1814 he almost immediately instituted a "close season for game" and "rewards for those who destroyed such noxious animals as 'tiger, [leopard] wolf, and wild cat, mouse-hound and hawk,'" each basically unenforceable, but soon thereafter, nevertheless, a prohibition on "killing game without a license" (Millar, *Plantagenet in South Africa*, 63).[8] Somerset's pedigree, so to speak, and his inclinations, in other words, are astonishingly at variance with the major events and thinking of his time, but this disjunction also suggests that the accident of his time and place in the Cape was also instrumental in the parallel developments of emerging but compromised human/civil rights and emerging but compromised animal protection/preservation provisions (Peires, "The British and the Cape," 472–480; Theal, *Records of the Cape Colony*, 11:207–208). From writing legislation to "protect" (some) animals in 1822 to exacerbating the conditions whereby the position of "Protector of Slaves" in 1826 became necessary, Somerset stands in the center of the vast compass of British colonial concerns that allowed discrete categories to coexist as though their proximities were both coincidental and mutually fortuitous: an ethical continuum, as it were, of the humanism du jour that meant the coextension of custodianship over *some* humans and *all* animals—two nascent and parallel systems of institutional segregation or apartheid, in other words.[9]

Thus, it was not until the passage of Ordinance 19 in 1826 (after Somerset's departure), the year in which *Unconfessed* opens, that laws governing slavery came into line with more liberal and workable pre-emancipation legislation elsewhere in the empire, particularly the Caribbean. This established the post of guardian of slaves (or "Protector," in the 1831 Consolidated Ordinance), whose authority to uphold the new provisions profoundly unsettled the stranglehold masters had had over their slaves' lives. The superintendent's visit, which opens *Unconfessed*, is fraught with what Sila recognizes as both his ignorance of and inability to change the circumstances over which he now wields the power—but not the knowledge—to do so:

> The guards and the field cornets, the police, the *landdrosts* [district administrators], the court clerks, the fiscal [prosecutor], the judges, everyone was going to teach him, this man, just how things were done here. She knew this, and she knew that what she had seen when he stood inside her cell was the first sign of a thread working its way loose around a button. Yes. They would try and undo the very secrets of life that held this man together. (6).

Sila is correct in part because Somerset's Proclamation of 1823 had indeed attempted to forestall reform in the name of reform. In addition to abolition of the slave trade in 1807–1808 and the subsequent proclamations such as those mentioned above, this Christian-centric proclamation had stoked the "mistaken belief" that it "had somehow freed [the slaves and] had allegedly sparked a minor, but bloody, and, to the masters, very frightening, revolt in the Koue Bokkeveld,"[10] a revolt that characters in *Unconfessed* respond to.

Concurrently, within twenty years of the novel's historical action, recreational "big-game" hunting in South Africa had, despite Somerset's protective legislation, bagged unprecedented, staggering, numbers of trophies for the first generation of "white hunters" on the continent. Accounts and memoirs of hunters' experiences rapidly became bestsellers, tapping into the appetite for "wilderness" adventures that were at once lurid, sentimental, and grotesque.[11] The grotesque in particular characterized the increasing conflation between the crossover universes of hunting, sport, game, and slavery/Africans. In a seminal essay, "With Camera and Gun in Southern Africa: Inventing the Image of Bushmen, c. 1880–1935," Paul S. Landau offers a devastating history and analysis of the parallel development between the uses of photography and the gun in the creation of the category "bushmen" initially *as* (nonhuman) animals. Tracing the mechanics of "*imagining*" to "actual picturing," Landau shows how "bushmen" shifted from "appearing bestial and depraved, to become gentle 'harmless people'" through "the connection between photography and naturalists' valorization of wild animals" (129). The meld of media in creating this colonially discursive web of its subject peoples is now well documented, but rather than simply (if chillingly) analogous and metaphorical, the characterizations were actually interchangeable, offering *no* distinction: "Some of them . . . began increasingly to hunt and kill bushmen themselves with firearms [c. 1900]"; "bushmen were cast as parasites, feeding off the mobile edge of civilization, much as if they were animal 'vermin'," and "the same people who would be shot with cameras were shot with rifles" (131–133). Similarly, as John MacKenzie reminds us, deep into the twentieth century Baden-Powell, the founder of the Boy Scouts, also noted with pleasure and nostalgia "the chase for 'wild beasts of the human kind,'" and "the longest march seems short when one is hunting game . . . lion or leopard, boar or buck, nigger or nothing" ("Chivalry, Social Darwinism, and Ritualized Killing, 51–52). Variations of this macabre theme continued with the formation of national parks in East Africa, where "'Primitive' Africans were often simply regarded as fauna" and that today still animate the "performance" of the "primitive" to entertain tourists (Neumann, *Imposing Wilderness*, 128).

Additionally, as these and other historians of empire and conservation have noted, Somerset's game laws, ineffective though they were at that time, nevertheless are also among the foundations of British colonial *land policies*, through the formation of game reserves and national parks during the nineteenth and twentieth centuries.[12] As such, then, wildlife, as objects of sport, sustenance, export, commerce, and recreation through tourism, has served as a multiple material base at several levels of the Cape Colony and South African economy, allowing those in control to appropriate land in the name of animal-based dominant human interests as well as generate revenue in the name of conservation (eventually) and species protection. These are all aspects of development that are both colonial and global. Colonial not only because of the timeframe but because these financial exploitations of "natural resources" systematically pitted wildlife, in this case, against multiple subordinated (human) group interests, both through land dispossession and, in the case of hunting, curtailed livelihood activities. Global because the multinational flows of capital operate in much the same way now with the wildlife tourist industry remaining as or even more important to the postapartheid neoliberal economy than before.

This becomes clearer in my other bird story, which, along with its tropes, involves a bird *in translation* that helps fulfill Sila's prescient and proleptic vision about big-game or wildlife tourism on the continent, where a not-so-subtle shift in value from some humans, in this case still "Africans," to animals and the environment has been fully instituted. The game of birds in *Unconfessed* confirms a network of signification that bypasses and undermines the "mute" status of subaltern knowledge. Conversely, in the Grzimeks' documentary *Serengeti Shall Not Die*, the consolidation of big-game and wildlife politics *reestablishes* a network of antihuman (African) justifications, but in the name of "conservation." Rather than a resource, such as the ostrich feathers of the late nineteenth century, or Sila's evocation of gulls as ubiquitous and loud record keepers, an additional level of a *naturalization* of birds occurs within the world of twentieth-century conservation, this time through the additional agency of material technology, namely, aircraft, photography, and film all as mechanical prostheses rather than as figurative objects alone.[13]

Mocking Who?

In 1961, Bernhard and Michael Grzimeks' documentary film *Serengeti Shall Not Die* won an Oscar. While the film is an ethnography of their census of the wildlife in Serengeti in northwestern Tanganyika, its primary purpose was to publicize the need to save the area for conservation, primarily of wildlife—sensitivity to ecosystems and biodiversity was to come later. Michael Grzimek had already enjoyed unprecedented success in being the first to use mass media—film and television programming—for advocacy and huge fundraising within Germany for wildlife causes in eastern and central Africa. His earlier film, *No Room for Wild Animals* (1956), filmed in the Congo, had amassed something of a fortune, allowing the Grzimeks the ability to propose purchasing land to help expand the Serengeti. The colonial government declined the offer and instead invited the two to help conduct the census, which was to determine if an expansion or partitioning was merited.[14]

Science as a credential, however, was a necessary baseline for this project, since it was up against the politics of an almost independent nation bent on both modernization and the development of its *human* capital. Rather than methodology or mathematical paradigms, the sign of science had to be *technology*—aircraft, photography, film, surveillance—materialized and plain to see, mainly because the environmental/ecological science itself was clearly wobbly. Furthermore, surveillance was another base function because the animals' migratory paths were crucial to the argument that a huge portion of the region be kept human (as in African) free. (The area has been described in various incarnations as the size of Connecticut, Northern Ireland, or Kuwait.) The effect sought by the father-son team needed to be more than simply comprehensive; it needed to appear all-encompassing, omniscient. God and "science" can thus be understood as the film's frame, as its Derridean *pareregon*,[15] where constructing perceptions about conservation were as important and enduring as the wildlife census (which was wildly inaccurate, as it turns out). In a somewhat parallel argument, Tobias Boes argues that the film pioneered an early version of "orbital photography," with its "visual experiments" that help evoke a "planetary consciousness" whose framework is one of a "natural heritage," ultimately turning the Serengeti into a (protected) "heritage site." So while the aerial photography is guilty of employing an "imperial gaze," Boes continues, it also compounded this ability with manifold manipulations of "scopic control" that subsumed the particular—Tanganyika—into the universal—for "*all mankind*" (48–53, my italics).

The Grzimeks continued to achieve such effects through complex sets of bird games. Where the novel plays its game of birds as a critique of colonial capitalist formations, the film plays games of birds that are integral to the realization of a fully fledged neocolonial set of formations manifest in wildlife conservation: its logic of conserving animals and their alleged pristine "wilderness" from the encroachment of humans, a form of separate development. In addition to its visual rhetoric, its "text," the film's games are also played within the apparatus and technology of animal surveillance, inclusive of tracking, hunting, and capture. The Grzimeks' original method was to employ the camera in a pioneering use of the bird's-eye view, through the platform of the aircraft. When that proved unfeasible, they maintained the vantage of the aircraft but employed their own eyes—like birds. Accordingly, the materiality of the film's understanding of "nature," including wildlife and birds, both contradicts and subverts that very materiality. An entanglement of "nature," language, and technology, therefore, operates within the film's practice, particularly in its cinematic manipulations and limitations of point of view: the combined techniques of lens disposition/foci and camera angle/location.

What cannot be underestimated on a material level is that there was no template for a task of the magnitude the Grzimeks had agreed to undertake, though the proximity to World War II and its advances in filming from the air and using the air for surveillance served as a backdrop.[16] They had to devise and invent methods as they went along. This included developing what became the tranquilizer gun to sedate animals, for example, as well as a pole lasso subsequently used to capture animals for zoos, and, postproduction, the manufacture of the modern "package tour." But the most time-consuming and challenging task remained finding a reliable, indeed, scientifically sound method of conducting the census. The aircraft was obviously one device, as was photography. The actual method for counting, however, posed more complex challenges.

The original plan of using aerial photography to conduct the animal-census count, following the logic of aerial photography pioneered during the war, was quickly abandoned when it became apparent that distinguishing one species from another was impossible from the altitudes aircraft usually need to maintain, and the expense of producing the necessary quantity of photographs to blanket the area under scrutiny was prohibitive. Rather than photography, then, the Grzimeks developed a method of training themselves and their assistants to lock in visually to groups at a time and therefore count in blocks or "groups of a dozen to fifty head," as they were flown up and down areas of approximate grids, also measured by the pilot's naked eye.[17] While this was an early and crude—possibly also the first—example of the area-multiplied-by-density approach, the substitution of the naked eye for the camera and any stereoscopic equivalent would appear to reverse the technological advantage that the camera and aircraft signified, and indeed it did. On the one hand, it implicitly reinforced the discredited "naked-eye science" associated with the photography and film sponsored by Carl Akeley and the American Museum of Natural History in the early twentieth century, which collapsed any distinction between what the eye and the camera "see," making the camera an instrument of "science" (Haraway, *Primate Visions*, 42–46). On the other hand, since that time, practices of aerial photography through the Second World War had long learned that in order to "read" and resolve accurately in 3-D the images taken from aircraft, one's eye had to be trained in elaborate adjustments and distortions for the hyperstereoscopic instruments to be used effectively, in order to compensate for the necessarily flattened and, in wartime, camouflaged topography photographed from above.

The implications, here, then, have less to do with the Grzimeks' method's accuracy of count, which was later proved significantly inadequate anyway, and more with the curious transfer of technological power from the machine (back) to the (white) men. According to Paul Saint-Amour, whose account I am heavily paraphrasing, these images created using hyperstereoscopic devices were "as if through eyes set farther apart than their own . . . [as if] seeing through the eyes of a giant" ("Modernist Reconnaissance," 360). The Grzimeks' naked (birds') eyes, then, borrow this specter of gigantism even while bypassing its technological agent, thereby enhancing not only the massive paternalism in play but also evoking the supernatural, an odd corollary to the prosthetic bind of the film's technology. The method displaces technology by investing the naked eye with its powers because it can, not because it makes scientific sense. Cost, more than anything, forces a recourse to ingenuity, which then becomes the surreptitious power underwriting the whole enterprise and a sign of and further confirmation of "white" superiority.

It is useful to foreground these technical details about the inability of unaided and unmediated aerial photography—particularly the vertical shot—to depict anything but a centrally flattened and peripherally distorted set of images, along with the naked eye's inability to discern animal species from a certain altitude. This reveals, first, a correlation between a fiercely naïve and positivist realism and the putative, even ad hoc "scientific" method at the base of the Grzimeks' project and film. It provides, in addition, a lesson in wide-angle photography and its own politics of representation: the camouflage of the politics of race played out in this expatriate scientific expertise, with its (wink) naked eye.[18]

Within this visual rhetoric, one of the first games of bird in the film is illustrated in part by Tobias Boes in his close reading of an early sequence of animated and then aerial footage that is used to introduce the first sight of Serengeti in the film. Commenting on the vantage of the small plane, Grzimek the elder says that "one sees something of the earth, just like a migratory bird," employing an idiomatic adjustment in the original German whereby, Boes explains, "earth" is substituted for "world," cleverly highlighting "the physical 'planet' that underlies the conceptual invention of any 'world.'" "By flying in a small plane," Boes continues, "we become like migratory birds, attuned to the landscape in a way unlike that ordinarily available to human beings" (51). This sensation is not only maintained in the opening sequences but heightened when a "'bird's eye perspective'—a zoomorphic view" of the park is "solidified" when the plane, its sound cut from the soundtrack, and its speed decelerated, appears to join a vast flock of rising flamingoes: "The illusion is one of disembodied weightlessness." The credits on the screen then associate with Michael's eventual fatal accident in that same aircraft, when the plane collided with a vulture during a similar maneuver, transmogrifying the "zoomorphic view" into "the beatific vision of a dead man" or "the perspective of a disincarnated subject" (51–52). With the vulture, the perspective merges birds, specters, and gods.

The film scholar and theorist Akira Mizuta Lippit illustrates particularly well these very hinges between animals, film, technology, and specters, which share a far more complex series of relations than as either subject/object or metaphor. His *Electric Animal: Toward a Rhetoric of Wildlife* (2000) opens with a sweeping truism, via John Berger, surely plucked out of the conservation debates schematized above and still propagated today. Lippit describes how animals have necessarily helped constitute modernity through their "disappearance":

> No longer a sign of nature's abundance, animals now inspire a sense of panic for the earth's dwindling resources. Spectral animals recede into the shadows of human consumption and environmental destruction. With the prosperity of human civilization and global colonization, ecospheres are vanishing, species are moving toward extinction, and the environment is sinking, one is told, into a state of uninhabitability.

"Arguably," he continues, "modernity has cost existence its diversity, has strained the earth's capacity to maintain life. It is a cliché of modernity: human advancement always coincides with a recession of nature and its figures—wildlife, wilderness, human nature, and so forth" (*Electric Animal*, 1).

Not surprisingly, Lippit's is a *sly* rendering of the cliché, even while echoing so closely the Grzimeks' view. At the same time, he recognizes that animals become "spectral" within this tableau and that they help metonymize "nature" as one of its "figures," along with, among others, "human nature." He goes on to note that within "the new economy of being during the modern period . . . animals never *entirely* vanish. Rather, they exist in a state of *perpetual vanishing*" and that through "modern technological media generally and the cinema more specifically, [they remain] *spectral*" (1). In response to the question "Why are animals the ideal subjects of photographs?" Lippit goes on to explain that "Without belonging to any ontological category animals have made those categories possible by situating their borders. It is only in the imaginary of the photograph that one is able, perhaps, to perceive or discern the animal world. It is in such an exposure that the animal enters, for the first time, the phenomenal world" (184). As such, then, animals and film, moving and still, help orchestrate what Lippit calls "*the transfer of animals from nature to technology*" (24, my emphasis).

The suspension, as in maintenance, of "perpetual vanishing" and "the spectral" here is akin to Boes's transmogrification of the zoomorphic view into that of the "disincarnated subject," as well as the "herds as a previously invisible conceptual entity," as each are subsumed into and out of the technical, manipulative, and illusory capability of film (Boes, "Political Animals," 47). But rather than merely optical and metaphysical illusion, at least another layer of transmutation, or another politics, is operational within the Grzimeks' film. Toward the end of the film, for example, a curious image of convergence or consolidation takes place that offers an extraordinary insight into the film's more surreptitious process of quite literally naturalizing its technology and its nature. The segment has made reference to the second aerial operation that has had to be conducted to gather information and samples of the flora and grasses of the park and, in doing so, bolsters its argument for the park's integrity and expansion. While the voiceover intones on the urgency of expanding the borders, inside and outside shots intercut to show the plane landing amid a flock of vultures disturbed by the aircraft. The plane then taxis around and next to a dead zebra, the object of the vultures' attention, and, as its propeller stops rotating, a similarly painted zebra-striped Land Rover drives up and parks between the plane and the dead animal. For an instant, a dead zebra and a Land Rover and an aircraft both painted as zebras fill the frame. Seamlessly, Michael gets out of the plane and climbs into the Land Rover, and the vehicle drives off, leaving the empty plane and dead zebra in the frame.

The scene is a brief segue into footage of a pride of lions arriving to feed on the dead zebra and then rest in the shade of the stationary aircraft.

Quite aside from the scene's work in the voiceover's narrative, the two mechanized zebras alongside a real but dead one suggests that their joint operation achieves a dual surveillance not only of "poachers' activities" but also of "Nature" in its cycles of wildlife predation. Thus, their convergence also acknowledges a shift in balance between the mechanized prostheses and the animals, where the vehicles, not the animals (or vehicles *as* animals), maintain the so-called natural order. This of course underwrites the ideology of the entire enterprise that the film is a part of: the silent marker of what, prepositionally and materially speaking, maintains "conservation." Modernity, then, metonymized as "technology" in zebra-painted vehicles, camouflaging not only itself but also the politics of its human agency, the surrogacy of which is in plain view even while disavowing its difference—in other words, the play of black and white, itself metonymic of the preternaturally colonial order of things. These visual relays of camouflage and bold statement as feints of perception in the larger enterprise of the film's "science" suggest yet another transference. The visually literal becoming animal (Deleuze and Guattari notwithstanding) between the human, nonhuman (spectral) animal, and the machines is more than simply instrumental. Rather than, or in addition to, the memory of the dead Michael's aura guiding a subtextual and parahuman point of view, the zebra carcass is spectral *not* because it is dead but because it is used to justify the larger operation through its visual, evidentiary, ocular proof of poaching as one of the gravest threats to wildlife in the area. The indictment-cum-justification itself joins a long colonial tradition of first demonizing, then criminalizing, and finally banning local, traditional African hunting practices, in all their variety, throughout the continent.[19]

This brings us around again to Sila as the mute subaltern, in a novel where the coded, figurative, and antitheory game of the birds of theory pinpoints both colonial biopolitics in deep operational mode and the (violently) epistemic moment of a specific branch of colonial capitalism. Christiansë's Sila is not tied to any piety of the sanctity of life, engaging in another palliative murder later in the novel; rather, she challenges the terms *within* which she is confined, the moral economy of a partially "ameliorated" slavery, one that supposedly anticipates "liberation" or "freedom." In the world within which the Grzimek film was made, the (human) subalterns are most immediately a group of Ikona and other groups from the western Serengeti and Maasai, severally displaced from the Serengeti and later handicapped within the Ngorongoro even while, in the latter case, the newly formed Ngorongoro Conservation Area (NCA) and its leadership, the Ngorongoro Conservation Area Authority (NCAA), purported to accommodate their continued residence.[20] Sila is able to pitch her recognition of colonial nonsense with the (anti)theoretical rejoinder of postcolonial shit; some of the Maasai can too, outside the realm of fiction but within a spectral discursivity. They are caught between the new urgencies of a brilliantly universalized conservation imperative and its dual mandate with "modernity," in the guise of development taking the form of a tourist/conservation industrial complex.

The historical context is dense. It was no coincidence that a major International Union for the Conservation of Nature and Natural Resources (IUCN) conference was held in Arusha, Tanganyika—the gateway to the Serengeti—on the eve of the nation's independence in 1961. Enjoying the spotlight of the Grzimeks' film's international acclaim, participants included the World Wildlife Fund (WWF) and other zoological and conservation organizations. One outcome of the conference was a document known as the Arusha Manifesto, which was released under the signature of the man who was to become the nation's first premier, Julius K. Nyerere. The statement outlined and sanctioned an extraordinary relationship between the new nation, its wildlife, and international nongovernmental organizations.[21] As Raymond Bonner discovered, however, this is a startling moment of pure (post)colonial ventriloquism: The document was not the work of Nyerere at all but rather the brainchild of E. M. Nicholson, dubbed by UNESCO's first head, Sir Julius Huxley (himself an avid naturalist, trained biologist, brother to Aldous, *and* a fervent eugenicist), to organize a fundraising arm for the IUCN (among whose founders in 1948 Nicholson had been), which was to become the WWF. The document, then, Bonner reports, was "written by Europeans, including Nicholson and Ian MacPhail, an advertising executive hired by WWF; they even titled it the Arusha Manifesto" (Bonner, *At the Hand of Man*, 63–65). In *Imposing Wilderness*, his exemplary study of the formation of the Arusha National Park, Roderick P. Neumann notes that "Conservationists were delighted by the manifesto, and it continues to be cited in their documents and publications as a positive example of African government interest and cooperation in protecting wildlife" (140).

In the interests of "modernization" and development, the Nyerere government both in its first decade and deep into its second, which included its socialist years, continued to subscribe to the pledge. The governing park authorities continued to circumscribe western Serengeti people and Maasai mobility and freedom of domicile within the parks' areas, and by

1975, the Maasai were evicted for good by the postcolonial state. This was a classic neocolonial move later developed into a postsocialist neoliberal stand that continues, even today, to maintain the original status quo.[22] So, still struggling two decades later, the displaced but remaining Maasai participated in the production of the video entitled *Enkigwana-ee-Ramat (Voices from Ngorongoro alt. Meeting of the Caretakers)* (1996). Created literally to insert their hitherto excluded and thwarted voices into the NCAA deliberations on the 1994 General Management Plan (GMP), the video was sent to the authorities, only to be ignored, as was an offer to have them respond in a like manner. Without reception and without reciprocity, the prosthetic potential of video was easily disabled. After this episode, which angered the IUCN, the NCAA banned the videoing of all such traditional local meetings within the NCA (Hodgson, *Being Maasai*, 93–95).

Nevertheless, the video is of a series of traditional community meetings and speaks eloquently out of its immediate context. After an introduction by Francis Ole Ikayo, in which he explains the timing of the recordings and the intended audience, each subsequent speaker is introduced, usually with a shot of a map marking his or her village, the date of the recording, and the speaker's name and status. The camera almost exclusively is trained upon the speaker, who is either in a classroom setting or outside with a gathering. Occasionally, while the speaker speaks, the camera pans the audience and in one episode zooms in and out of a long shot of the village environs. Third parties come and go across the screen, including, in one instance, perhaps, the videographer, since it appears to be a white man. Elders and leaders of all the affected Maasai and related communities displaced from both the Serengeti and later Ngorongoro (NCA) address either their constituencies or the NCAA via the camera.

Speaking Maa, with English subtitles, each explain either their duped participation and subsequent withdrawal or the underhanded tactics used by the government and park agencies (NCAA) to extract agreement and claim compliance for their displacement and/or the curtailment of their access to supplemental cultivation. In a series of scathing speeches, we hear sophisticated analyses of how illiteracy, language differences, and time restrictions have been tactically used in an elaborate plan of disenfranchisement. While the elders and leaders acknowledge the terms of the GMP, which includes the prioritized needs of wildlife, food security via government subsidies rather than supplemental cultivation, and deceptive promises of some infrastructure, they also provide a historical context. A young warrior (no other name is given) starts with, "Here's what we say about the plan: when it came we said it's singing a song to celebrate our demise. We say it's like condolences, because they have finished us." He later claims: "Since cultivation was banned they [NCAA] haven't known what we've been eating. They only care what the wildlife eat. . . . If they say the NCAA is internationally valuable—are we not part of it? Are we not part of that international thing?" He ends by observing: "Their voices [of the educated Maasai] are not heard beyond the NCA because there are many others as far away as America, whose children are fed from this land's value."

The warrior's words, along with the whole video, helpfully complicate the conditions of possibility of "salvage," as well as notions of the "spectral," even as he challenges the still operational denial of coevalness for his group within this still ongoing negotiation.[23] He even evokes the rhetoric of "all of mankind." Politically, is this subaltern mute or spectral? The film and video replicate the conditions of impossibility in the proxy from colonial to neocolonial, from nonsense to more non-sense. The postcolonial shit, here, is as sharply and pungently accurate as was that of Sila's birds of (anti)theory.[24]

Mocking Bird!

The continued game of birds in *Serengeti Shall Not Die* confirms this, over and over. The Grzimeks' employment of aerial photography and the operational bird's-eye view, including its wide-angle lens, offer more than just metaphor, analogy, and *becoming*; instead, this bird-game also includes what Derrida names an *animot* and Akira Lippit an *animetaphor*. *Animot* is cognizant of the *misuse* of animality in figurative language and jars spoken French as a "chimerical word" (Derrida, *The Animal That Therefore I Am*, 41). Drawing on Derrida, Akira Lippit in *Electric Animal* outlines what an *animetaphor* is in this way:

> One finds a fantastic transversality at work between the animal and the metaphor—the animal is already a metaphor, the metaphor an animal. Together they transport to language, breathe into language, the vitality of another life, another expression: animal and metaphor, a metaphor made flesh, a living metaphor that is by definition not a metaphor, antimetaphor—animetaphor. The animetaphor may also be seen as the unconscious of language, of *logos*. (162)

In addition to the zoomorphic implications and transmogrifications of their bird's eyes and points of view, a further wrinkle emerges in the film when we learn that in Swahili, the word for aircraft is *ndege*, bird. What emerges if we continue to read this as an *animot*/*animetaphor* is that, in addition to the aircraft as a photographic and conservationists' platform allowing for a "bird's-eye view," which is both a pre- and post-technological (ani)metaphor (that also carries the narratological vantage of oversight and omniscience), the "bird's-eye view," obviously, also plays on the notion of the "perspective of the bird," or "zoomorphic view," evoking, perhaps, a metaphysical "bird's point of view." The "bird" in question, however, is the mechanized one, and the "subjectivity" animated by the act of conservation (surveillance) occurs through the power of proxy, *on the birds'* (wildlife's) *behalf.* The posthumanist implications are striking not because of its anomalous status during that period, since the Grzimeks' advocacy had already made itself known through their robust anthropomorphic zoocentrism, but in its structural entrenchment *as* apparatus. In this film's agonistic equation between humans and nonhuman animals, therefore, the antagonist is not (really) "humans" but "Africans" (and, more particularly, western Serengeti groups and Maasai), keeping locked in place one of colonialism's most enduring segregations. The proxy representation here masks a *political* rather than *scientific* operation.

Finally, by looking again, the aircraft, a single-engine Dornier DO-27, is *painted to look like a zebra.* The problem here is that *ndege* mangles the visual camouflage of a flying zebra by now making zebras birds; in other words, it again achieves the status of *chimera*, a true *animot.*[25] The bizarre along with the phantasmagoric is wholly in line with the unlikely actuality of a mute (rather than muted) subaltern. Within the annals of conservation, therefore, this ability to so distort threads a fine irony, given that *Serengeti Shall Not Die* became instrumental in ultimately putting into place the model known as "fortress conservation."[26] As its name suggests, this was the practice of segregating humans and wildlife through the gazetting of national game parks, where human habitation was banned. By creating this segregation, or separate development, a corollary of the Cartesian human/animal split, the myth of saving a "wilderness" was maintained even while large-scale dispossession and relocation of (some) human populations took place, assuring generations of conflict, as we witness in *Enkigwana-ee-Ramat.* In creating vast swathes of African-free land, however, *Serengeti Shall Not Die* also laid the groundwork for the park's repopulation, but not only by wildlife but also by expatriate scientists, foreign tourists, and hunters (and perhaps *some* Africans, but as service providers, only). Sila's birds of (anti)theory continue to keep an eye on and disentangle the inner but open secret of colonial and neocolonial capitalism. Postcolonial shit and colonial nonsense are still locked in an agonistic struggle.

NOTES

1. For exhaustive historical background and analyses on archival work for *Unconfessed*, see Christiansë, "'Heartsore.'"

2. Christiansë, "'Heartsore,'" 12–15, offers a concise digest of the ongoing controversy over infanticide in the colony at the time of the novel's action. For further analyses of infanticide in South African slavery, see Van der Spuy, "Infanticide, Slavery, and the Politics of Reproduction," 128–148; and Scully, *Liberating the Family*, 143–146. For archival references to infanticide, see Theal, *Records of the Cape Colony*, 11:176–183, 344–349. As always, I'm indebted to Yvette Christiansë for sharing her research with me.

3. See Theal, *Records of the Cape Colony*, 25:150–155. Game laws had existed since the Dutch issued some in the mid–seventeenth century distinguishing between "protected animals" and "vermin." See MacKenzie, *The Empire of Nature*, 202n2. Somerset's proclamation falls into a second wave of game laws within the colony more in line with the interests of an "imperial elite." See the anti-MacKenzian essay of Van Sittert, "Bringing in the Wild, 273."

4. See Derrida, *The Animal That Therefore I Am*, 82, where he characterizes the Cartesian "theoretical animal," or "the animal of theory," as "seen" rather than "sees." My claim is that Christiansë's birds of (anti)theory not only see but see through.

5. My equation is obviously selective, but several formulations of the cognates suggest somewhat similar outcomes if we always keep in mind that, as human waste, slavery is always someone else's gain.

6. I'm grateful to Chris GoGwilt for bringing Cadava's "The Guano of History" to my attention.

7. "Animal turn" is a coinage of Harriet Ritvo ("On the Animal Turn," 118–122) that she uses to explain "new relationships between scholars and their subjects, and new understandings of the role of animals in the past and the present" within the academy.

8. See MacKenzie, *The Empire of Nature*, 203; Adams and McShane, *The Myth of Wild Africa*, 45; and Brown, "Cultural Constructions of the Wild," 84–85, who also cites prior Dutch efforts to regulate hunting, dating back to 1669, whereas MacKenzie cites 1657–1658 and 1684. See also Beinart, "The Night of the Jackal," 197. For the entire 1822 Proclamation, see Theal, *Records of the Cape Colony*, 25:150–155.

9. I'm making a broad generalization here. The actual categorization of wildlife into different subgroups was a complex taxonomy in itself and reflected both the commodification of some wildlife as "vermin" feeding a bounty economy, others as "royal game" for hunting/sporting purposes and protection therein, and "to reserve the wild population [of Ostrich] for

domestication . . . transforming them into private property." See Van Sittert, "Bringing in the Wild," 271–275.

10. See Mason, "The Slaves and Their Protectors," 108; Rayner, "Wine and Slaves," 161–166; and Scully, *Liberating the Family*, 38–45.

11. See Ritvo, *The Animal Estate*, Part 3, "Animals and Empire," 205–288.

12. See McKenzie, *The Empire on Nature*, 203; McShane and Adams, *The Myth of Wild Africa*, 45; and Beinart and Hughes, *Environment and Empire*, 60, who see the region as "a touchstone in the empire for preservation and conservation." See also Brown, "Cultural Constructions of the Wild," 80–81.

13. See Van Sittert, "Bringing in the Wild," 274–278, for a detailed account of the mid–nineteenth century boom in ostrich feathers and the processes of domestication they underwent.

14. For an accounting of many of the Grzimeks' tactics and their endeavors, as well as for alternative readings of the film, see Tobias Boes, "Political Animals," 41–59; Laken, "*Serengeti Shall Not Die*," 224–264; Mitman, *Reel Nature*; Hayes, *The Last Place on Earth*; Neumann, "Ways of Seeing Africa," 149–169; and Shelter, *Imagining Serengeti*. For a detailed overview of the partitioning of Ngorongoro, see Rogers, "History and Governance in the Ngorongoro Conservation Area," 78–117. I'm indebted to an anonymous reader for alerting me to the Boes article.

15. See Derrida, *The Truth in Painting*.

16. See Boes ("Political Animals") for a reference to a German-language film article (Hediger, "Das Tier auf unserer Seite") that confirms that the Grzimeks' film "draws on the conventions of military photography and Nazi cinema to assert control over a landscape even as it purports to merely describe it" (Boes, "Political Animals," 43). Boes also draws from Claudia Sewig's biography of Grzimek, which documents his friendship and film apprenticeship with Leni Riefenstahl. See Sewig, *Bernhard Grzimek*. Accordingly, it's worth noting that Leni Riefenstahl's opening footage from an aircraft, including aerial shots of marching troops on the ground with the shadow of the aircraft traversing their ranks in *Triumph of the Will* (1935), was among the first uses of the aircraft as camera platform as well as a deifying symbolic incorporation of point of view. See Boes, "Political Animals," 50; and Evans, "Air War, Propaganda, and Woolf's Anti-Tyranny Aesthetic," 59–60. I'm grateful to an anonymous reader for reminding me of the Riefenstahl usage. The same visual metaphor, of the shadow of the aircraft shot from the aircraft itself—shaped like a cross—moving over the ground/water, is a device used extensively in Hubert Sauper's eastern- and central African–based documentaries, such as *Kisangani Diary* (1998), *Darwin's Nightmare* (2005), and, more recently, *We Come as Friends* (2014).

17. See the book that documents the film venture, Grzimek and Grzimek, *Serengeti Shall Not Die*, 65; and Hayes, "The Last Place," 78–79. Boes offers a different reading of this circumstance, suggesting instead that shots from inside the cockpit showing "park rangers assisting the Grzimeks track population figures through hash marks on note pads [allows the viewer to] participate in a quite literal fashion in the 'discovery' of the East African herds as a previously invisible conceptual entity" ("Political Animals," 47).

18. This does not preclude the film's ability to dazzle and awe with its sweeping panoramic shots of vast herds on the move, what Boes calls "breathtaking sequences shot from within the cockpit that give aerial views of the stampeding herds" ("Political Animals," 47).

19. See MacKenzie, *The Empire of Nature*, 201–224; Neumann, "Dukes, Earls, and Ersatz Edens." *Serengeti Shall Not Die* has its own sequence on poaching, where, in an obviously staged reenactment, the plane is used to apprehend poachers ("human hyenas") using tactics reminiscent of World War II air warfare, such as strafing. As Boes, quoting Hediger, notes, the camera was often placed where a machine gun would be mounted on a warplane, which Michael famously evokes in a quotation from the book: "I'd like to have a machine-gun that fired through the propeller," referring to "poachers" (qtd. in Boes, "Political Animals," 47; Grzimek and Grzimek, *Serengeti Shall Not Die*, 181).

20. For a full accounting, see Shivji and Kapinga, *Maasai Rights in Ngorongoro*; Neumann, "Ways of Seeing Africa," 149–169; Shetler, *Imagining Serengeti*, 1–4, 135–237; and Rogers, "History and Governance."

21. The statement, the "manifesto," reads in full: "The survival of our wildlife is a matter of grave concern to all of us in Africa. These wild creatures and the wild places they inhabit are not only important as a source of wonder and inspiration but are an integral part of our natural resources and of our future livelihood and well-being. In accepting the trusteeship of our wildlife we solemnly declare that we will do everything in our power to make sure that our children's grandchildren will be able to enjoy this rich and precious inheritance. The conservation of wildlife and wild places calls for specialist knowledge, trained manpower and money and we look to other nations to co-operate in this important task—the success or failure of which not only affect the Continent of Africa but the rest of the world as well. Julius K. Nyerere (7.9.61)." See MacKenzie, *The Empire of Nature*, 324–325; and Adams and McShane, *The Myth of Wild Africa*, 113–114. The manifesto should not be confused with the 1967 Arusha Declaration, which outlined the nation's socialist agenda.

22. This push for "modernization" took absurd but grim forms, including attempts to ban Maasai traditional garb, which was deemed indecent,

"primitive," and "backward." For a detailed analysis of this and other such bans, see Ivaska, *Cultured States*, 37–85; and Hodgson, *Being Maasai*, 66.

23. I'm evoking Fatima Tobing Rony's definition of "salvage enthnography," whereby film was used as a scientific tool to "capture," and thus "save," so-called vanishing tribes in the early twentieth century. See Rony, *The Third Eye*. Haraway, in *Primate Visions*, earlier coined a similar phenomenon, "celluloid taxidermy."

24. Hodgson's study covers the travails of the groups affected by wildlife legislation among other factors in northern Tanzania, primarily but not exclusively the Maasai from the colonial period to 2010 or thereabouts. It includes the array of strategic political frameworks inclusive of forming multiple NGOs adopted and abandoned to effect somehow the broader enfranchisement within the postcolonial state. The state in almost all its incarnations—liberal, socialist, postsocialist, and neoliberal—has persistently denied full participation, initially in the interests of "modernity" but more fully in the interests of resource and land appropriation, exploitation, and development. One strategy employed by the Maasai included winning the status as "Indigenous" recognized by the United Nations in the 1990s. Into the new millennium a shift to the framework of "pastoralist livelihoods" from "Indigenous" and the creation of Civil Society Organizations (CSOs) and community-based organizations (CBOs) rather than just foreign donor–sponsored NGOs took place. Hunter-gatherers have not enjoyed the same level of advocacy, alas, signaling another dismaying wrinkle in the politics of disenfranchised groups within Tanzania, namely, the fetishized (as exotic and or/ "indigenous") status of "Maasai" by international misreadings at the expense of other pastoralists such as the Barabaig or other nonpastoralists such as the Hadzabe. Today, the sophistication and exploitation of digital media are now standard tools for many of the pastoralist advocacy groups, even while the fact of pastoralism itself continues to be eroded.

25. The Swahili name for zebra is *punda milia*, which translates as striped donkey.

26. For a detailed case history of this method of conservation, see Brockington, *Fortress Conservation*. Since the advent of structural-adjustment implementations and other neoliberal policies since the 1990s, a variety of different conservation models have emerged. "Community conservation," in particular, is an ongoing phenomenon and has evolved in myriad ways through the region and the continent, with varying degrees of success. It purports to share dividends from wildlife tourism with the local communities, either through leasing or partnerships. In almost all its incarnations, the balance of power remains with the nonlocal players. See Neumann,

"Primitive Ideas"; Hulme and Murphree, eds., *African Wildlife and Livelihoods*; Walpole and Thouless, "Increasing the Value of Wildlife through Nonconsumptive Use"; and two notable documentaries, Simpson, *Milking the Rhino*; and Apostolidis, *A Place without People*.

WORKS CITED

Adams, Jonathan S., and Thomas O. McShane. *The Myth of Wild Africa: Conservation without Illusion*. Berkeley: University of California Press, 1992.

Apostolidis, Andreas., dir. *A Place without People*. Anemon Productions, 2009. DVD.

Beinart, William. "The Night of the Jackal: Sheep, Pastures, and Predators in the Cape." *Past and Present* 158 (February 1998): 172–206.

Beinart, William, and Lotte Hughes. *Environment and Empire*. Oxford: Oxford University Press, 2007.

Bhabha, Homi, K. "Articulating the Archaic: Cultural Difference and Colonial Nonsense." In *The Location of Culture*, 123–138. London: Routledge, 1994.

Boes, Tobias. "Political Animals: *Serengeti Shall Not Die* and the Cultural Heritage of Mankind." *German Studies Review* 36 (2013): 41–59.

Bonner, Raymond. *At the Hand of Man: Peril and Hope for Africa's Wildlife*. New York: Knopf, 1993.

Brockington, Dan. *Fortress Conservation: The Preservation of the Mkomazi Game Reserve, Tanzania*. Bloomington: Indiana University Press, 2002.

Brown, Karen. "Cultural Constructions of the Wild: The Rhetoric and Practice of Wildlife Conservation in the Cape Colony at the Turn of the Twentieth Century." *South African Historical Journal* 47 (November 2002): 75–95.

Cadava, Eduardo. "The Guano of History." In *The Other Emerson*, ed. Branka Arsic and Cary Wolfe, 101–129. Minneapolis: University of Minnesota Press, 2010.

Christiansë, Yvette. "'Heartsore': The Melancholy Archive of Cape Colony Slavery." *Rewriting Dispersal: Africana Gender Studies* 7, no. 2 (Spring 2009). http://www.barnard.edu/sfonline/africana/print_christianse.htm.

———. *Unconfessed*. New York: Other Press, 2006.

Derrida, Jacques. *The Animal That Therefore I Am*. New York: Fordham University Press, 2008.

———. *The Truth in Painting*. Trans. Geoff Bennington and Ian McLeod. Chicago: University of Chicago Press, 1987.

Evans, Elizabeth F. "Air War, Propaganda, and Woolf's Anti-Tyranny Aesthetic." *Modern Fiction Studies* 59, no. 1 (2013): 53–82.

Grzimek, Bernhard, and Michael Grzimek. *Serengeti Shall Not Die*. Intro. Alan Moorhead. Trans. E. L. Rewald and D. Rewald. New York: Dutton, 1961.

Grzimek, Michael, and Alan Root. *Serengeti Shall Not Die* and *No Room for Wild Animals* [*Serengeti darf nicht sterben*, 1958; *Kein Platz für wilde Tieire*, 1956]. Allied Artists Pictures, 1959. DVD.

Haraway, Donna. *Primate Visions: Gender, Race, and Nature in the World of Modern Science*. New York: Routledge, 1989.

Hartman, Saidiya. *Lose Your Mother: A Journey along the Atlantic Slave Route*. New York: FSG, 2006.

Hayes, Harold T. P. "The Last Place." *New Yorker*, December 6, 1976.

———. *The Last Place on Earth*. New York: Stein and Day, 1997.

Hediger, Vinzenz. "Das Tier auf unserer Seite: Zur Politik des Filmtiers am Beispiel von Serengeti darf nicht sterben." In *Politische Zoologie*, ed. Anne von Heiden and Joseph Vogl, 287–301. Berlin: Diaphenes, 2007.

Hodgson, Dorothy L. *Being Maasai, Becoming Indigenous: Postcolonial Politics in a Neoliberal World*. Bloomington: Indiana University Press, 2011.

Hulme, David, and Marshall Murphree, eds. *African Wildlife and Livelihoods: The Promise and Performance of Community Conservation*. Oxford: James Currey, Heinemann, 2001.

Ivaska, Andrew. *Cultured States: Youth, Gender, and Modern Style in 1960s Dar es Salaam*. Durham, N.C.: Duke University Press, 2011.

Laken, Thomas. "*Serengeti Shall Not Die*: Bernhardt Grzimek, Wildlife Film, and the Making of a Tourist Landscape in East Africa." *German History* 29, no. 2 (2011): 224–264.

Landau, Paul S. "With Camera and Gun in Southern Africa: Inventing the Image of Bushmen: c. 1880–1935." In *Miscast: Negotiating the Presence of Bushmen*, ed. Pippa Skotnes, 129–141. Cape Town: Cape Town University Press, 1996.

Lippit, Akira Mizuta. "The Death of an Animal." *Film Quarterly* 56, no. 1 (Fall 2002): 9–22.

———. *Electric Animal: Toward a Rhetoric of Wildlife*. Minneapolis: University of Minnesota Press, 2000.

MacKenzie, John M. "Chivalry, Social Darwinism, and Ritualized Killing: The Hunting Ethos in Central Africa up to 1914." In *Conservation in Africa: People, Polices, and Practice*, ed. David Anderson and Richard Grove, 41–61. Cambridge: Cambridge University Press, 1987.

———. *The Empire of Nature: Hunting, Conservation, and British Imperialism*. Manchester: Manchester University Press, 1988.

Mason, John Edwin. "Hendrick Albertus and His Ex-Slave Mey: A Drama in Three Acts." *Journal of African History* 31, no. 3 (1990): 423–445.

———. "The Slaves and Their Protectors: Reforming Resistance in a Slave Society, the Cape Colony, 1826–1834." *Journal of Southern African Studies* 17, no. 1 (March 1991): 103–128.

Millar, Anthony Kendal. *Plantagenet in South Africa: Lord Charles Somerset.* Cape Town: Oxford University Press, 1965.

Mitman, Gregg. *Reel Nature: America's Romance with Wildlife Film.* Cambridge, Mass.: Harvard University Press, 1999.

Morris, Rosalind. "In the Name of Trauma: Notes on Testimony, Truth Telling, and the Secret of Literature in South Africa." *Comparative Literature Studies* 48, no. 3 (2011): 388–416.

Nabizadeh, Laila. "A Duplicitous Dialectic: Hegel's Master-Slave Dialectic in *Unconfessed* and *This Earth of Mankind*." Unpublished undergraduate comparative literature paper, Fordham University, 2012.

Neumann, Roderick P. "Dukes, Earls, and Ersatz Edens: Aristocratic Nature Preservationists in Colonial Africa." *Society and Space* 14, no. 1 (1996): 79–98.

———. *Imposing Wilderness: Struggles over Livelihood and Nature Preservation in Africa.* Berkeley: University of California Press, 1998.

———. "Primitive Ideas: Protected Area Buffer Zones and the Politics of Land in Africa." *Development and Change* 28 (1997): 559–582.

———. "Ways of Seeing Africa: Colonial Recasting of African Society and Landscape in Serengeti National Park." *Ecumene* 2, no. 2 (1995): 149–169.

Peires, J. P. "The British and the Cape, 1814–1834." In *The Shaping of South African Society, 1652–1840*, ed. Richard Elphick and Hermann Giliomee, 472–518. Middletown, Conn.: Wesleyan University Press, 1989.

Rayner, Mary. "Wine and Slaves: The Failures of an Export Economy and the Ending of Slavery in the Cape Colony, South Africa, 1806–1834." Ph.D. diss., Duke University, 1986.

Ritvo, Harriet. *The Animal Estate: The English and Other Creatures in the Victorian Age.* Cambridge, Mass.: Harvard University Press, 1987.

———. "On the Animal Turn." *Daedalus* 136, no. 4 (Fall 2007): 118–122.

Rogers, Peter J. "History and Governance in the Ngorongoro Conservation Area, Tanzania, 1959–1966." *Global Environment* 4 (2009): 78–117.

Rony, Fatimah Tobing. *The Third Eye: Race, Cinema, and Ethnographic Spectacle.* Durham, N.C.: Duke University Press, 1996.

Saint-Amour, Paul. "Modernist Reconnaissance." *Modernism/modernity* 10, no. 2 (2003): 349–380.

Scully, Pamela. *Liberating the Family? Gender and British Slave Emancipation in the Rural Western Cape, South Africa, 1823–1853.* Oxford: Heinemann, 1997.

Sewig, Claudia. *Bernhard Grzimek: Der Mann, der die Tiere liebte.* Bergisch GladbachL Lübbe Verlag, 2009.

Shelter, Jan Bender. *Imagining Serengeti: A History of Landscape Memory in Tanzania from Earliest Times to the Present.* Athens: Ohio University Press, 2007.

Shivji, Issa, and Wilbert B. Kapinga. *Maasai Rights in Ngorongoro, Tanzania.* London: IIED/HAKIARDHI, 1988.

Simpson, David E., dir. *Milking the Rhino.* Written by and photography by Jason Longo. Kartemquin Film and Independent Lens, 2008. DVD.

Theal, McCall. *Records of the Cape Colony.* Vol. 11: *November 1815–1818.* 1902.

———. *Records of the Cape Colony.* Vol. 25: *January 1st to 6th February 1826.* 1905.

Van der Spuy, Patricia. "Infanticide, Slavery, and the Politics of Reproduction at Cape Colony, South Africa, in the 1820s." In *Infanticide: Historical Perspectives on Child Murder and Concealment, 1550–2000*, ed. Mark Jackson, 128–148. Aldershot: Ashgate, 2002.

Van Sittert, Lance. "Bringing in the Wild: The Commodification of Wild Animals in the Cape Colony/Province c. 1850–1950." *Journal of African History* 46 (2005): 269–291.

Walpole, Matthew J., and Chris R. Thouless. "Increasing the Value of Wildlife through Nonconsumptive Use? Deconstructing the Myths of Ecotourism and Community-Based Tourism in the Tropics." In *People and Wildlife: Conflict or Coexistence?*, ed. Rosie Woodroffe et al., 122–139. Conservation Biology 9. Cambridge: Cambridge University Press, 2005.

Weaver, John. "Litigating for Freedom in the British Empire, 1815–1822: The Universal and Local in Tension." In *Empire and Autonomy: Moments in the History of Globalization.* Vancouver: University of British Columbia Press, 2009.

CHAPTER 9

Of Mimicry, Birds, and Words: The Technology of Starling Song in European, American, and Indonesian Poetry

Christopher GoGwilt

sukanta's	stevens's	schubert's
jalak	skreaching	miller
bali	grackle	made
speaking	graham's	a
only	skittering	starling
itself	starlings	sing
the	flock	europe's
starling	to	lyric
song	america	heart

"An eye for resemblances"—this is the sign of "genius" for Aristotle (*On the Art of Poetry*, 31). Walter Benjamin elaborates on this "gift" at the beginning of his short essay "On the Mimetic Faculty" ("Über das mimetische Vermögen"): "Nature creates similarities. One need only think of mimicry. The highest capacity for producing similarities, however, is man's. His gift of seeing resemblances is nothing other than a rudiment of the powerful compulsion in former times to become and behave like something else" (*Reflections*, 333).[1] "An eye for resemblances"—this "gift" that Benjamin and Aristotle single out as the distinctively human capacity, a particular feature of the human faculty of language—appears also to be the gift humans characteristically ascribe to starlings, parrots, and other mocking birds. In their mimicry of human speech, mocking birds have repeatedly promised poets and scientists alike to show and tell what is (and is not) distinctively human about language.

Here I want to examine the part that the starling plays in this more general economy of mocking bird mimicry of language and mimesis: in the animal mimicry of man, in what birds tell us about human language, and in the mimetic faculty of (human-animal) mocking bird technologies. The starling—as bird and word, zoologically and tropologically speaking, as it appears poetically and biopolitically—is an especially revealing kind of mocking bird. What it resembles, how it resembles, and the way it appears and disappears (individually, in flocks, as a species, in its relation to other kinds and other species, and in the part it plays across a range of scientific and nonscientific, poetic and nonpoetic texts)—these are the questions I want to address. Focusing in particular on the appearance of the starling in poetry from across the world—Europe, North America, and Indonesia—I want to explore what the technology of starling song tells us about the faculty of language and, beyond (or alongside) human language, the "mimetic faculty."

Charles Darwin provides a classic point of reference in *The Descent of Man*, referring to "parrots and other birds" for their ability to see the way language connects "words with things, and persons with events" (219). Darwin's comments are especially interesting because they are embedded within an analogy between the evolution of language as traced by comparative philology and the evolution of species as traced by himself, outlining a close relation between evolutionary theory and comparative philology.[2] The starling is only one of those "other birds" (besides parrots) that offer special insight into the "distinguishing character" of human articulation. The relative insignificance of the starling in relation to "parrots," however, suggests a comparative contrast between different kinds of birds, and this may have implications for the evolution of species, for the evolution of language, and for an understanding of the "mimetic faculty" in the most general of aesthetic, biological, or ecological senses.

For Darwin, "parrots and other birds" are invoked to illustrate the mimicry of *sound*, whereas the sense of mimicry privileged in the citations from Aristotle and Benjamin is that of *sight*. Both senses, however, are intertwined. Indeed, the coordination of different kinds of sense-perception may be one of the riddling effects presented by the particular example of the starling, a bird whose stereotyped appearance is split between its parrot-like facility for imitating human speech and its striking visual appearance (the iridescent sparkle of its glossy feathers or the acrobatic displays of whole "murmurations" of starlings in flight). Neither the auditory nor the visual sense can be entirely disentangled from all of those other senses activated by these stereotypes—something emphasized by Tim Birkhead's *Bird Sense* in its human attempt to imagine "what it's like to be a bird."[3] We should pay closer attention to the overestimation of auditory and visual senses (and the interrelation between each form of the overvaluing of one sense) in order to assess the multiplicity of senses implicated in the complex behavioral and poetic example set by the starling. The linguistic significance of starlings' characteristic ability to imitate sounds (human and nonhuman) may perhaps depend as much on nonlinguistic behavioral features, such as those that enable starlings to coordinate the stunning flight patterns for which they are equally as famous. The English term "murmuration" is revealing in signifying both the visual acrobatics *and* the auditory effect of starling flocks.[4] That overlap of senses involved in the coordination (production and reception) of auditory and visual registers already points to a simultaneously *linguistic* and *paralinguistic* function. This may be crucially determining for the so-called language faculty that scientists and linguists continue to debate (turning, frequently, to starlings for scientific examples).[5]

This double sense of mimicry (auditory and visual, linguistic and paralinguistic) is replicated in the different emphases placed on the mimicry of "parrots and other birds" by Darwin and Alfred Russel Wallace, the two European father figures of evolutionary theory. A. R. Wallace emphasizes the visual form of "mimicry" (protective coloring) in his contributions to the theory of natural selection, whereas Darwin emphasizes the auditory form in his discussion of language in *The Descent of Man*. The seemingly minor place of the starling in Darwin's and Wallace's accounts of animal mimicry, moreover, is potentially revealing for showing how this double register of mimicry implicates theories of language and evolution in the kind of human mimicry Homi Bhabha has explored as the ambivalent logic of colonial discourse ("Of Mimicry and Man"). Darwin and Wallace both seem to reproduce a distinction between the parrot and the starling that might crudely be put in this way: The parrot is the global sign of what may be found everywhere in the differentiation of species; the starling is a particular European commonplace instance. Wallace's passing formulation "our own starlings," shifting from auditory to visual registers, suggests a casual distinction between European and extra-European species lodged at the heart of evolutionary theory: "The talking Mynahs, like our own starlings, build in holes, the glossy starlings of the East (of the genus *Calornis*) form a hanging covered nest, while the genus *Sturnopastor* builds in a hollow tree" (376). Here, as elsewhere in Wallace, "our own starlings" provide a European point of reference for what occurs in much more interesting variations elsewhere. The place of the starling in Darwin's text is arguably indifferent to the distinction between parrots and starlings. It is relegated to a note in which several instances of parrots' use of language come before a single instance of a starling speaking stock German phrases ("good morning" and "good bye, old fellow"). This very indifference, however, suggests a kind of global positioning of the starling as an effect of commonplace European habits mimicked throughout the world. If the parrot is a visibly global sign of the ambivalence of colonial mimicry, the starling secures a European commonplace signifier for one of those "other birds" that parrot. The very ordinariness of the "starling" (and the word itself) reproduces an effect of variation—camouflage, mimicry, and differentiation—that ultimately eludes, even as it orchestrates, evolutionary classification.

In an effort to undo the constraints of evolutionary classification, Deleuze and Guattari in *A Thousand Plateaus* open a question of "becoming-animal" that continues to serve as a point of reference for debates about the relation between the human and the animal.[6] They situate their argument against the combined wisdom of classical theories of mimesis, romantic theories of language, and evolutionary theories of natural selection—namely, "the mimetic or mimological vision [that] made the idea of an evolution-production possible" (234–235). By contrast, their "molecular" vision suggests a radically different kind of understanding of the technology of birdsong: "no longer the songbird, but the sound molecule" (248). Seeking to foreground the "sound molecule" in the technology of birdsong, Deleuze and Guattari nonetheless emphasize that the problem of mimesis or "the mimetic," "mimological vision," "is in no way behind us" (235). Already implied by the "refrain" of birdsong and songbirds throughout *A Thousand Plateaus*, the recurrent problem of language and mimicry (either in the classical sense attributed to Aristotle's *mimesis* or in the archaic sense attributed by Benjamin's "mimetic faculty") might simultaneously be approached as a "molecular" question of "assemblages" in the Deleuzian and Guattarian sense. The theoretical problem of language (especially the presupposition of the distinctively human faculty of language) demands attention to the implication of technology (and technological assemblages) in multiple (auditory and visual) senses. In both Aristotle and Benjamin the overestimation of *the visual* attends a theoretical grasp of language more usually focused on *the auditory* (speech; *phonè* and *logos*). They seem thus to situate the whole question of *mimesis*, the mimetic, and representation to the side of language, suggesting that there is something besides linguistic phenomena (some combination of faculties, techniques, or technologies) that makes for the distinction accorded both to language and to the human. It is this paralinguistic element of mimesis—the molecule in birdsong—the starling may help identify. Starling behavior may lead science and poetics alike to appreciate a "mimetic faculty" above and beyond the "faculty of language."

The assemblage of technologies specific to all the various "starlings" referred to by Darwin and Wallace might provisionally be called the technology of starling song (what is alluded to in Wallace's formulation "the talking Mynahs" and linked, too, to the characteristic way starlings mimic sounds from their surrounding ecosystem). I propose three hypotheses about this starling-song technology. The first hypothesis is that the starling song is a technology that repeatedly crosses over species: not only different kinds of starlings (starlings and mynahs), not only different birds (Darwin's "parrots and other birds"), but also different kinds of animal (birds and humans). The second hypothesis is that this technology of starling song tracks these transspecies effects across a global (which is also to say comparative) history of language (the "faculty of language" for Darwin and recent scientists such as Chomsky, Hauser, and Fitch). The third hypothesis is that this global history (of species and language) emerges in paralinguistic effects—in between the sensory, linguistic, and ecological systems of birds and humans—that might be called starling tracking effects (and with an eye to considering the combined auditory and visual registers of such effects). Such effects may point toward that kind of "ancient reading" Walter Benjamin describes at the end of "On the Mimetic Faculty"—"reading before all languages, from the entrails, the stars, or dances" ("das Lesen vor aller Sprache, aus den Eingeweiden, den Sternen oder Tänzen" [336; 94]). This archaic, prehistoric, and prelinguistic sense of reading might stand as a definition of the paralinguistic register tracked by starling effects.

Schubert's "Ungeduld," from Die Schöne Müllerin

First, consider the starling at the heart of European lyric form, as it appears in the seventh song of Franz Schubert's *Die Schöne Müllerin* (1824), "Ungeduld" (Impatience), based on the poem in Wilhelm Müller's 1820 cycle of poems Schubert chose for his musical setting. Here the lover imagines training a young starling ("einen jungen Star") to speak "with the sound of my voice" ("mit meines Mundes Klang") the words "yours is my heart" ("Dein ist mein Herz"). The romantic irony with which the human voice is made dependent on the mimicry of the bird may be just another variation on the staple trope of birdsong as emblematic of lyric form. Yet by contrast either to the generic bird or to that most emblematic of songbirds, the nightingale, the starling inverts the staple feature of that lyric trope—rather than having the human voice imitate the bird's song, the bird is made to imitate the human voice. This inversion itself might be traced back to what Sarah Kay has examined as the interrelation between two different lyric paths at the origins of European lyric poetry—the "nightingales' way" and the "parrots' way" as these emerge in troubadour poetry, the first embodying the originality and creativity of a "natural" mastery of song, the second embodying the "fundamental human capacity for mimesis."[7] Linked to the practice of lyric quotation in troubadour poetry, the "parrots' way," according to Kay's account, loses ground to the "nightingales' way" as the dominant, determining feature of European lyric poetry.

The particular starling effect in Schubert's "Ungeduld," however, is also linked to a whole constellation of technologies producing the romantic irony of its German lyric form: The miller's song is constantly reproducing the sound of the wheels of the mill in the water of the brook, constantly turning pastoral idyll into suicidal compulsion. The whole apparatus of modern technology is recapitulated in the effect of this lyric voice, whether read in the original Müller song or in Schubert's musical setting—or even later in Liszt's transcription(s) of that song for solo piano. The lyric voice is an effect of this setting, itself uniquely trained—like the "young starling"—to articulate the repeat performance of its own aesthetic education. It is a repeat performance that compulsively repeats the violence done to the world as the violence done to itself, undone again in the merely mechanical repetition of the words of another: "yours is my heart and so it should always be" ("Dein ist mein Herz, und soll es ewig bleiben!").

This violent commonplace at the heart of European lyric form (and as the apparatus of so-called Western European modernity) is both generic—albeit generic as a transspecies effect of voicing (echoing, naturally and literarily, the Narcissus myth)—and also quite specifically the technology of a starling's ability (or compulsion) selectively to imitate sounds from its surrounding ecosystem.

All three of our hypotheses might be tested out in relation to this one example. First Hypothesis: The technology of starling form is transspecies—the human voice depends on the "mimetic faculty" of the starling to articulate itself. Moreover, the longer history of lyric form within which this vocal training is inscribed suggests a perpetual crossing over of different species and kinds ("mimicry" associating and differentiating starlings, parrots, and other mocking birds; "mimicry" also differentiating different kinds within a species—"our starlings," "the talking Mynahs," "the glossy starlings of the East"—not to mention different genders, ostensibly the underlying form of resemblance determining the compulsion to mimic in the first place, or perhaps in the last instance).

Here, again, we might turn to Sarah Kay's argument, noting the exemplary if also exceptional function of Marcabru's "starling" (*estornel*) songs in defining the difference between the parrots' way and the nightingales' way.[8]

Second Hypothesis: The technology of the starling tracks a global history that is undecidedly both European and non-European. It is not clear whether the starling of Schubert's Müller song is chosen for the specific place and name of its German setting or whether its European difference from other mocking birds is already inscribed in the thought of its vocal training. Kay's argument is especially illuminating here, suggesting that this very hesitation or undecidability is inscribed into the structure of European lyric form (and as itself premised on "a divergence in the reception of troubadour poetry between northern France and Empire on the one hand . . . and Catalonia and northern Italy on the other" [23]), and again Marcabru provides a revealing point of reference, looking back to classical examples like Ovid, looking sideways to the debatable interrelation with Arabic and Persian traditions, and looking forward to the consolidation of European lyric form (around the "nightingales' way") with Dante and Petrarch. This "starling" mimicry at the heart of European lyric form tracks a hesitation in any study of mimesis considered in comparative multispecies, geocultural, and linguistic perspective.

Third Hypothesis: In making the formative element of human linguistic development hinge on paralinguistic mimicry (the whole apparatus of the watermill mimicked in the starling's mechanical replication of the human voice), this starling effect repositions the comparative study of language, the aesthetic education of humankind, and the biopolitical apparatus of modern technology in the most general of senses within a global constellation of different and overlapping systems of both nature and culture.[9]

Starlings in Jorie Graham and Wallace Stevens

My second example dwells on the reference to starlings in Jorie Graham's poem "The Dream of the Unified Field" (which gives the title to her *Selected Poems* from 2002—originally the poem appeared in the 1995 volume *Materialism*). In the second section of the poem, the speaker's unresolved account of relation to her daughter (from the first section) is set in counterpoint to the description of a snowstorm in which a flock of starlings appears, a description that combines visual and auditory effects into a striking moment of lyrical expansion:

Starting home I heard—bothering, lifting, then bothering again—
the huge flock of starlings massed over our neighborhood
these days; heard them lift and
swim overhead through the falling snow
as though the austerity of a true, cold thing, a verity,
the black bits of their thousands of bodies swarming then settling
overhead. I stopped.

(176–177)

The spectacle of this "huge flock of starlings" (notably evoked both in its auditory and visual registers: "I heard—bothering, lifting, then bothering again") arrests the homeward movement of the speaker to yield a lyric experience—a vision, approaching something like an epiphany—that turns the starlings' appearance almost into a universal romantic symbol, although the lyrical expansion is traced along its limits at the same time:

Foliage of the tree of the world's waiting.
Of having waited a long time and still having
to wait. Of trailing and screaming.
Of engulfed readjustments. Of blackness redisappearing into
downdrafts of snow. Of indifference. Of indifferent reappearings.

(177)

These last lines before the speaker returns to her thoughts of "you" (her daughter from the first section) seem to enact a movement that simultaneously evokes and evades making the trope of a starling flock into a symbol of purgatorial "waiting" (with a possible reference to Dante's starling trope from Canto 5 of the *Inferno*) or something more natural-supernatural in the romantic tradition (for example, Coleridge's description of a murmuration of starlings discussed in Gavin Sourgen's chapter in this volume). Whatever one reads into these starling precursors, the effect of Graham's starling trope is to show "the way things work" (to quote from the very first poem in the *Collected Poems*). The lyrical "I" insists on the combination of auditory and visual—"trailing and screaming"—that draws out and unravels a series of starling analogies, notably, "Of blackness redisappearing" and (opening and closing the movement toward romantic symbolization) "Of indifferent reappearings."

What makes the specificity of the starling trope all the more interesting is the way it is modulated in the third section of the poem. The starling trope is simultaneously held onto and let go as the lyrical "I" turns to "the single cry / of the crow" (whose auditory and visual imagery will come to dominate the fourth section of the poem)—

Then I heard it, inside the swarm, the single cry

of the crow. One syllable—one—inside the screeching and the skittering,
inside the constant repatterning of a thing not nervous yet not ever
still—but not uncertain—without obedience
yet not without law—one syllable—
black, shiny, twirling on its single stem,
rooting, one foot on the earth,
twisting and twisting—

These lines (and their shifting from the first one-line stanza to the second seven-line stanza) both recall and seem to forget again the trope of the starling in a movement that reduplicates the outward lyrical movement of the second section in the allusion to the "swarm"—more specifically "inside the swarm." Here, Graham's poem calls our attention to the double valence of auditory and visual imagery in the starling trope: the impossibility of separating the two classificatory features of the starling, its sound and its movement ("trailing and screaming"), even as she refuses, or evades, the term "murmuration." The term she uses instead—"swarm"—ties sound and sight together but appropriates the trope of the starling and subsumes it into a rather different—wider, animalistic-human—vocabulary, at the same time incorporating it into her own poetic idiom (recalling her collection *Swarm*).

Incorporated into Graham's poetic idiom, the starling trope of this section is also an embedded citation of Wallace Stevens—specifically, in its allusion to the "skreaking and skrittering residuum" of Wallace Stevens's grackle in "Autumn Refrain." As the third section of Graham's poem turns away from the opening evocation of the "huge flock of starlings" in the second section, the "screeching and the skittering" of the "swarm" develops what might be read as a retrospective reading of the trope of the starling in Stevens's poetry. The grackle is Stevens's starling—by analogy, but also, zoologically speaking, a distant relation, a New World species. "Autumn Refrain" contrasts the "grackle" with the "nightingale" to figure the idiom of American poetry in counterpoint to that of the European tradition. Citing Stevens's "skreaking and skrittering" grackle—but turning it ever so slightly toward the different auditory and visual register of "the screeching and the skittering"—Graham alludes to that poem's "evasions of the nightingale" (taking us back, not coincidentally, to that much longer, if also more complex, literary history of the European lyric outlined by Sarah Kay in terms of the counterpoint between nightingale and parrot, between the practices of lyric insertion and lyric quotation). The specificity with which her own poetic idiom has absorbed the technology of starling mimicry and is in the process of forgetting it in turn reveals in retrospect the extent to which Stevens's own poetic idiom (and its claim on an American poetic idiom and on modernist poetics more generally) also turns on the technology of starling form.

If the grackle, for Stevens, is a form of starling—the New World counterpart to the European starling—its technology turns on an irreducibly ambiguous effect of reference. Its "evasions of the nightingale" are not simply, as Harold Bloom has put it, a modernist troping of the romantic poetics of Keats's nightingale.[10] There is (as Hollander's elaboration on Bloom's point makes clearer) a vertiginous unraveling of grounds of reference for what kind of poetic birdsong is evoked by its "skreaking and skittering residuum." The mimetic and antimimetic play famously captured in Stevens's modernist aesthetics turns simultaneously on what we might call the "grackle" effect of pure mimesis (the onomatopoeia of imitating the sharp screech of the grackle's ugly, very un-nightingale bird call) and the "grackle" effect of pure antimimesis (the semiotic play of "skreaking and skrittering" as sounds that imitate the mocking sound of "grackle" as a name for a bird). Given the fact that grackles can also mimic the sounds of other birds—making them a sort of lesser-order mocking bird (a mock mocking bird: the "grackle" effect is, then, a mock mocking bird effect)—Stevens's grackle embeds in his poetics something like the same reversal of birdsong we saw at work in Schubert's song and also, at least potentially, the same paralinguistic effect of crossing over from linguistic to nonlinguistic features of the surrounding ecosystem.

The difference, for Stevens, is that its "evasions of the nightingale" specify a comparative taxonomy of confusion in the identification of sounds in a biological, ecological, and poetic sense. Wallace Stevens's signature "grackle" effect—"the skreaking and skrittering residuum"—foregrounds the linguistic and paralinguistic effects of starling technology in Schubert's "Ungeduld" but elides the trope of the "starling" almost entirely by elevating its New World relative, the "grackle." The ambivalent question of mimicry foregrounded in the example of the Schubert song (does the "starling" domesticate the exotic parrot, or is it the parrot's European original?) becomes the hidden mechanism for controlling the ambivalent mimicry of colonial discourse.

The elision of the starling in Stevens—and the double register of the auditory and visual identification that marks its signature disappearing trick—is all the more remarkable when related to the trope of the "blackbird" in Stevens's poetry—notably, in "Thirteen Ways of Looking at a

Blackbird"—since the "blackbird," too, in Stevens, may be a "grackle," both mimetically speaking and as a poetic challenge to the very notion of mimesis. To the extent that the "blackbird" of Wallace Stevens's most celebrated poem (his signature poem, one might almost call it) is, as Bart Eeckhout has argued, "of a fundamentally *hybrid* species," its hybridity challenges the very notion of hybridity (colonial and postcolonial) on which classification systems depend.[11] Indeed, the specificity of the starling in its ambivalent absence from Wallace Stevens's poetry recalls the effects at work in the evolutionary theory of Darwin and Wallace. In both, the commonplace of the European "starling" (bird and word) reproduces an effect of variation—camouflage, mimicry, and differentiation—that ultimately eludes, even as it orchestrates, evolutionary classification.

The very absence of the starling in Wallace Stevens, then, returns us to our three hypotheses about the technology of starling song, specifying the classificatory problem of the European starling both in the texts of evolutionary theory and lodged at the heart of the formation of European lyric (as illustrated by Schubert). The ambiguity we found in Schubert—whether the "starling" (*Star*) replaces the contrast between nightingale and parrot or whether it already activates the dialectic of exotic and commonplace—is there in Stevens, too, albeit through an elision of reference to the starling itself (bird and word) that accentuates all the more the global, comparative history of "starling" technology. Its technology is all the more active for having been displaced (forgotten, or repressed) by the grackle and the blackbird.

It should be noted that this is simultaneously a matter of tropology and zoology, of poetic form and bio-/zoological fact, since the European starling had already been introduced into the American landscape in Wallace Stevens's lifetime. The history of the appearance of European starlings in North America offers a rather striking, crude example of the mixing up of poetry and ornithology, since the European starling was first brought to New York's Central Park in the early 1890s by the New York Zoological Society member Eugene Schieffelin, as part of his effort to introduce into North America all the birds that appear in the works of William Shakespeare.[12] Predating the publication of Stevens's poetry, the spread of European starlings throughout North America and the corresponding reshaping of the sounds and sights of its various ecosystems might be said to coincide with the emergence of Stevens's distinctive poetics. Although absent in the poems themselves, the European starling becomes a household name and sight around the same time Wallace Stevens appears as a distinctive American poet. In this sense, the absence of the starling in Stevens's poetry may be less of an absent presence than the effect of a poetic disappearing trick. The way Jorie Graham's poem tracks the effect of starling technology in her evocation of the American landscape suggests a correlation between the appearance of the European starling as a ubiquitous feature of the American landscape and the disappearance of the starling as trope (or rather, the technology of its disappearing act) in contemporary American poetics. In the poem that follows "The Dream" in the collection *Materialism*, her adaptation from Audubon's journal evokes a landscape *before* the appearance of starlings. Whether one emphasizes the antimimetic, negative dialectic of this linguistic disappearing act (whereby Stevens's poetry mimics but elides the European starling) or its reversal (whereby the appearance of the European starling as part of the North American ecosystem unconsciously shapes Stevens's poetics), the disappearing act of starling technology calls attention to the way the limitations of language (however defined—philosophically, scientifically, or according to the "language faculty") are framed by paralinguistic effects (both auditory and visual).

The disappearing act of starling technology in the poetics of Wallace Stevens is something Graham's poem goes on to explore in the third and final section of "The Dream of the Unified Field." The third section of the poem turns from the starling to the crow—"Then I heard it, inside the swarm, the single cry// of the crow." The difficulty of reading this turn from starling to crow is the difficulty of the poem itself—the problem of separating the individual poem as poem and the attendant critical problem of reading the borders of the poem (as suggested by the title's "unified field"—which then, too, becomes the title for the *Collected Poems*), including the "I" of the poet's voice (a problem reiterated elsewhere in the collection *Materialism* by the five different poems entitled "Notes on the Reality of the Self"). Leaving aside the particulars of the movement of this lyrical "I" from section to section of "The Dream" (recounting the memory of a mother-daughter relation whose movement from section to section emerges around the starling effects I've been tracking so far), in the final two sections the poem dissolves the boundaries of this lyrical "I," turning to the strategy of adapting quotations from other texts used elsewhere in the volume, here specifically drawing from the journals of Christopher Columbus to mark a self-reflective contemplation of the European colonization of America that leads into a quotation continued over into the next and final section of the poem:

its vast
white sleeping geography—mapped—

> not a lease—*possession*—"At the hour of vespers
> in a sudden blinding snow,
> they entered the harbor and he named it Puerto de

The caesura between sections visibly marks the problem of the poem's borders—an exercise in meditation on what Daniel Tiffany calls "lyric substance" or the materiality (or "materialism") of poetry. Tiffany's reading of this poem is itself revealing in its emphasis on the trope of the snowstorm as an embodiment of "materialism" (both lyric substance in Tiffany's sense and the title of Graham's volume)—a reading that is revealing, for our purposes, in its evasion of the residual effects of "starling" technology in orchestrating the poem's final turn to Columbus's discovery of America:

> The nebulosity of the material storm and the solidity of intellectual objects coincide because the blind, possessive, agglutinative mode of composition is the same in both cases. Ultimately, the storm that is reassembled in the mind reveals itself to be a "possession" of history, a "splinter colony" (86). Indeed, in a startling transformation, the snowstorm becomes the "vast white sleeping geography" of the "new world" discovered by Columbus (87)—the very substance of a unified field of matter, thought, language, and history. (Tiffany, *Infidel Poetics*, 55)

This reading enacts the disappearance of the starling tropes on which the movement of the poem depends,[13] even though the poem itself repeatedly marks and remarks on that disappearance: first, materially, as "Of blackness redisappearing into/downdrafts of snow" (177); then, in the movement from starling to crow; additionally, too, in the possible reference, in "splinter colony," to the appearance of the European starling in North America. Indeed, that "redisappearing" starling effect may be found, too, in the figure of the "storm" that begins the sixth section (and that unifies the poem according to Tiffany's reading)—"The storm: I close my eyes and,/standing in it, try to make it *mine*" (180). The final turn toward the words of Columbus's journal is set up by a parsing of the relation between an "outside" and an "inside," both of which are effects marked by the materiality of the starlings' "disappearing" figures—"outside, the talk-talk of the birds" and "inside, a splinter colony, new world, possession" (181).

In the shift to the last section of "The Dream," the poem performs a caesura in the borders of the poem. One way to explain this break would be to see how the lyrical "I" of this poem (and others in the volume, in Graham's poetry as a whole, and indeed throughout the history of poetry) is marked off against all those other discourses—like the words taken from the Columbus journal—adapted from other texts and presented as poem-assemblages. Is it too much to suggest an analogy between the form of this adaptation and assemblage of other bits of discourse and the technology of starling song? In the original collection in which "The Dream" appeared, the very next poem is one that markedly presents an American landscape from a time and space before the appearance of the European starling. This prose-poem adaptation "from Audubon's Missouri River Journals (an adaptation)" not only continues from "The Dream" 's Columbus adaptation the effect of breaking from the boundaries of the lyrical "I." It mirrors its own dissection in an attentive description of the auditory and visual field of an American human-animal-and-bird life in the process of being measured, classified, and systematically broken down and killed (as Audubon did to produce those striking, mimetic images of *The Birds of America*). Graham's "adaptation" marks a haunting absence of reference to the starlings, so central a trope in the previous poem. Another instance—as with the poetry of Wallace Stevens—of the disappearing trick of starling technology.

The trope of the starling in Jorie Graham's "Dream of a Unified Field"—and the technology of starling song it reveals at work in the *disappearance* of the starling both in Wallace Stevens and in the staging of her own lyrical "I"—reminds us, again, of our three hypotheses: first, in projecting a hidden comparative anatomy of different forms of birdsong mimicry (human imitations of birds, bird imitations of humans); second, in projecting a global imagining of linguistic and literary migration—the European starling that Graham reintroduces into Stevens's poetry as an emblem of the migration of an invasive European species into the Americas; and third, correlating that global positioning of language and material culture as a remnant of that "nonsensuous similarity" Benjamin posits in "the most ancient" form of reading ("reading before all languages, from the entrails, the stars, or dances"), that paralinguistic leftover in the "skreaching and skrittering residuum" of Stevens's grackle replicated in the "screeching and skittering" starlings forgotten, recalled, disappeared, or *redisappeared* in the "one syllable" of a crow.

Putu Oka Sukanta's Song of the Starling

My third example is from the poetry of the Indonesian poet, novelist, and AIDS activist Putu Oka Sukanta, who was imprisoned for ten years in Salemba prison in Central Java under the "New Order" regime of Suharto.

The New Order regime (overturning the Old Order of Sukarno) was established under the pretext of suppressing the communist coup attempt of September 30, 1965, leading to the mass arrest, imprisonment, and deportation of political prisoners and the massacre of many thousands more suspected communists or communist sympathizers. Putu Oka Sukanta is not the most prominent writer to have given voice to the oppression of this traumatic moment in Indonesia's recent history, although his novelistic memoir of the experience of imprisonment (a short excerpt of which was published in *Manoa* under the title "Leftover Soul") might now be counted as one of a set of testimonial points of reference for an experience whose full significance is still being reckoned with.[14] *The Song of the Starling: Poetry of Oppression* (*Tembang Jalak Bali: Puisi Tertindas*), published in a bilingual edition with translations by Keith Foulcher in 2000, provides something of a lyrical counterpoint to the work of the most prominent dissident writer, Pramoedya Ananta Toer, whose prison memoirs, *Nyanyi Sunyi Seorang Bisu*, were published in an English translation in 1999 under the title *The Mute's Soliloquy*. Putu Oka was imprisoned in Salemba prison just outside Indonesia's capital, Jakarta, while Pramoedya (after a short stint in Salemba) was deported to the infamous prison camps for political prisoners in exile on Buru Island. Both writers sought to give voice to an experience of oppression under a state of exception that has, as Giorgio Agamben famously argues in *Homo Sacer*, become the rule of modern times: "The camp . . . is the new biopolitical *nomos* of the planet" (176).

Putu Oka Sukanta's choice of the "starling" (*jalak*) in the title of his collection of the "poetry of oppression" provides a compressed, lyrical version of the dissident challenge to the New Order regime more famously represented by all the work (fiction, memoirs, testimonials) Pramoedya produced in the Buru Island prison camps. When, in the first section of his prison notes, *The Mute's Soliloquy*, Pramoedya asks "Is it possible to take from a man his right to speak to himself?" the question resonates both as a testimonial reflection on the way Pramoedya was deprived of just such a right to "speak to himself" and as a defiant question whose premise is rejected in the very asking of the question and in the testimonial form of *The Mute's Soliloquy* in which it appears. The disturbing ambiguity of this questioning of a fundamental human right to freedom of expression has its counterpoint in Putu Oka's choice of the "starling's song" as metaphor for the "mute soliloquy" of his own experience of internment in Salemba prison.

In what sense, though, does Putu Oka invoke "starling" song as a mimicry, a poetics, or a technology about the deprivation of human rights and the experience of imprisonment? Putu Oka Sukanta's "starling" song seems to unravel all three of the hypotheses I have been entertaining so far. In the poem entitled "pilihan" ("Choices"), the single reference to the starling in the first line of the last stanza isolates a particular kind of starling—"*jalak bali*," the Bali mynah, or starling—insisting that its song (*tembang*) tells only of itself (*menceritakan dirinya*—literally, tells stories only about itself): As Keith Foulcher translates it, "the song of the starling is only about himself" (130).

The first thing to note is that the particular "starling" in question here is not the common European starling its English translation might suggest but rather the Bali mynah, an endangered species of bird restricted to Bali, the only vertebrate endemic to that island, and a celebrated symbol of Bali (featured, for example, on the Indonesian 200 rupiah coin). The very particularity of this bird, then, sets it apart from the "talking Mynahs" (in Wallace's formulation) by which other starlings are classified. The song of the Bali mynah, by contrast to the European starlings of Schubert or Jorie Graham, but also by contrast to the literal and figurative mynah birds of all the various cultural traditions converging on Indonesia, does not seem to be associated with any variety of mocking bird technology. Its song is invoked in isolation from any kind of human communication. Its song, in this sense, is indeed "about itself."

Putu Oka Sukanta's choice of the "Jalak Bali" (echoing the "choices" of the poem's title) surely hinges on this particularity (and the poem's Indonesian title word, *pilihan*, conveys, too, the evolutionary sense of natural selection). The lyrical effect is, in some respects, all too easy to explain in terms of the metaphorical identification of poet and bird—both from Bali, both threatened with extinction, and for both of whom "song" reproduces only this emblematic endangerment of an isolated species. There is something enigmatic in the very simplicity and directness of this identification with the Bali starling that makes it the lyrical counterpart to Pramoedya's question about the fundamental human right to free speech. Whereas Pramoedya's question foregrounds an animalization of the human, the deprivation of a model of language, communication, and political rights that would distinguish human from animal, Putu Oka's starling foregrounds a humanization of the animal, a total identification of poet and bird premised on a metaphorical symbolization of the starling as a species on the verge of extinction, as emblematic of Bali, as the very embodiment of what Agamben would call "bare life" (*nuda vita*). Together, one might say, Pramoedya's notes from Buru and Putu Oka's lyric songs from Salemba prison constitute an extended meditation on Agamben's chiastic formulation: "The total humanization of the animal coincides with a total animalization of man" (*The Open*, 77) ("*L'umanizzazione integrale dell'animale coincide con una animalizzazione integrale dell'uomo*" [*L'aperto*, 80]). Read in this way, the technology of starling song we have been reading at the heart of European lyric and in the modernist formation of American poetics may be nothing other than what Agamben calls "the anthropological machine" of modernity (*The Open*, 33–38; *L'aperto*, 38–43).

The movement of the poem "Choices" confirms this reading, at the same time suggesting we need to reconsider the technology of starling "song" in comparative global, linguistic, and cultural perspective. "Choices" (*pilihan*) develops a short series of three reflections, in three stanzas, on the challenge of leaving behind a record of the poet's experience. The first two lines already mark the experience of exile, deprivation, and precarious communication registered by the whole collection of poems: "*kutulis jejak pada buku harian / supaya kelak engkau dapat memastikan*" (129)—in Foulcher's translation, "I write down the traces in a diary / so you can confirm it all one day" (130). It moves from hopeful anticipation in the first stanza—the hope that the writer's "traces" will be "confirm"ed by the addressee one day ("kelak"—literally, "later")—to fearful anticipation in the second stanza: "*takutku pada rindu seperti takut / dikejar kata akhir dari perhitunganmu*" ("my fear of longing is like the fear / of meeting up with your last words of assessment"). The hopeful form of communication in the first stanza is balanced against the fearful anticipation of a final calculation—both, nonetheless, depending on a model of communication between an "I" and "you" that then disappears in the final stanza presenting the "song of the starling" that "tells only of itself."

The appearance of Putu Oka's starling coincides, then, with the disappearance of the lyrical "I" and its address to the "you"—a naturalization, or animalization, of human language that might be perfectly romantic in its symbolization (and remains that in the coin's reproduction of the Bali mynah) were it not that the starling's particular classification (scientific and as stereotyped) emblematizes a species on the verge of extinction. Here, by contrast to our other examples, the starling seems to isolate language and species in a fateful enclosure of song, a bleak, though lyrical and beautiful singing to oneself—the counterpart, as noted above, to Pramoedya Ananta Toer's "mute's soliloquy."

As we consider how the final stanza further develops this figure of the "starling song," we should consider some of the implications of Putu Oka's use of this figure as the title for the whole collection of poems or, even more generally, as the subtitle puts it, the "poetry of oppression." It is the exceptional nature of the "Bali" starling or mynah that makes it the perfect choice both for the particular poem and for the collection as a whole. In drawing attention to this exception—reminding us forcefully of Agamben's argument about the logic of exception at work in the logic of the "anthropological machine"[15]—the poem's selection of a starling reveals, by its very contrast to the other kinds of starling (as a question of zoology, poetics, and language), what is at stake in the technology of starling song.

First, dwelling on the exceptional nature of this starling that does *not* seem to mimic humans or communicate at all across species, we might consider the poem's own call for readers nonetheless to "listen" to the starling—"listen / listen" ("*dengarlah / dengarlah*")—which leads to the concluding contrast between the Bali starling/mynah and the owl—"it's not the chattering of the owl / inviting death" / "*bukan celoteh burung hantu / mengundang maut*" (128/129). It is tempting to read in this contrast between the "Bali starling" and "the owl" a symbolic contrast between bare life and death, yet the poem's imperative to "listen" resists symbolization insofar as it foregrounds the song that sings itself—or tells stories only of itself—or, in Foulcher's translation, "is about himself." Foulcher's translation adds a gendered difference that suggests the poet's identification with a male bird and retrospectively reads the first two stanzas as the male poet imprisoned in Central Java addressing his wife back in Bali. This is not only a plausible translation but probably the most responsible translation, yet, as with the symbolism of Bali starling as endangered species and owl as harbinger of death, the terms of the poem itself resist its figurations—if the poet's call to "listen" is addressed to the "you" of the first and second stanzas, it reiterates the shift from hopeful to fearful communication (the first "listen," as it were, figuring the addressee's deciphering the "traces" in the writer's "diary"; the second "listen" figuring the addressee's "last words of assessment"), yet the very grounds of human communication have been taken away by the figure of the "starling," whose song concerns only itself.

The presuppositions of language demanded by and enacted in this poem might at first appear to stand in stark contrast to the technology of mimicry associated with European starlings and mynah birds—not to mention parrots and mocking birds of all kinds. The logic of what we have provisionally called Putu Oka Sukanta's own "mute soliloquy" suggests, indeed, that his choice of the starling turns on just this exclusion. For sure, the poem itself contrasts the starling with the owl. In what sense does the choice of "Bali starling" not also imply a contrast to all the other starlings of the world and so engage, in a kind of negative dialectic, with the comparative anatomy and poetics of that more general technology of starling song?

Our reading here depends, in the most basic sense, on a question of translation that goes to the heart of whatever we might want to call the technology of starling song. If Keith Foulcher's translation "song of the starling" elides the specific genus "Bali mynah," the more generic sense of "starling" (*jalak*) may be necessary to emphasize the "choices" of comparison brought together in the resonance of the Indonesian phrase. "Jalak" is the common word (across various mutually understandable Indonesian languages) for what might either be translated as "starling" or "mynah." The difficulty of choosing between "starling" and "mynah" as the most appropriate English word might be explained in terms of the difficulty of choosing between a commonplace European word and an exotic word of Asian origin and perhaps also the difficulty of choosing the word that least evokes the image of a bird that imitates human speech. The act of translation itself reproduces the poem's own work of selection—natural, linguistic, and poetic selection—in choosing the "Bali *jalak*" (both for the poem's "choices" and as emblematic of the "poetry of oppression" in the collection's title).

There is something analogous at work in the translation of *The Mute's Soliloquy*, the title of the single-volume English edition of Pramoedya Ananta Toer's two-volume set of notes from Buru Island prison camp, *Nyanyi Sunyi Seorang Bisu*. The Indonesian title means literally "the quiet/silent/lonely song of someone who is mute"—the first part, *Nyanyi Sunyi*, an echo of a 1937 collection of poetry by Amir Hamzah that marks a literary turning point, the bridge between older Malay traditions of poetry and a new idiom of modern literature written in the revolutionary new language *bahasa Indonesia*. The second part, *Seorang Bisu*—a mute person, also possibly a deaf-mute—suggests the silencing of Pramoedya's writing and also the partial deafness he suffered after being beaten when he was arrested. The English translation *The Mute's Soliloquy* eloquently captures the extended meditation throughout the notes on the implications of being deprived of the right to speak—a "man's right," as we noted above, "to speak to himself." More than merely the complement in human speech to Putu Oka Sukanta's starling song, Pramoedya's "mute soliloquy" emerges, also, as a problem of translation. If Putu Oka Sukanta's *Bali jalak* presents a difficulty in deciding how to translate the human classification of the particular endangered species of bird, Pramoedya's *Nyanyi sunyi seorang bisu* presents a difficulty in deciding how to translate the animal conditions to which the human voice has been reduced. In the original Indonesian version of the passage cited above—"Is it possible to take from a man his right to speak to himself?"—the "voice" invoked is repeatedly figured as the voice of child, a baby, a mere infant struggling to communicate:

> Following the events of 1965, I lost everything or, to be more accurate, all the illusions I had ever owned. I was a newborn child, outfitted with the only instrument a newly born babe finds necessary for life: a voice. [The Indonesian here doesn't yet use the term "voice" but rather emphasizes that he was reduced to the state of dependency of a newborn child—so the term "voice" here, evoking the *neediness* of the infant, is not yet a "voice" of human rights, indeed is explicitly figured as being *already* reduced, dispossessed, the way a newborn child is.] Thus like a child my only means of communication was my voice: my screams, cries, whispers, and yelps. (13)

It is this voice—a voice before language, before any conception of human rights—that is then invoked in the question (which one might then translate with a slightly different inflection—"And to be deprived of *even that mode of communication*, that mere, needy, new-born voice, ach—is it possible to take from a person even that most bare, reduced right to commune with oneself?")

Even more striking than the problem of translation posed by this passage—where the question of human rights begins to look much more like the kind of attention to "bare life" that Agamben sees at work in the "anthropological machine"—is the problem of translation posed by the passage from which the English translation takes the title "the mute's soliloquy." Here, first, is the English translation (the italicized section provides that sense of the title phrase "mute's soliloquy" that I have been reading as the complementary formulation to Putu Oka Sukanta's "song of the starling"), followed by the corresponding passage in the Indonesian edition (the italicized section—translated, at the end, into English—provides part of the section never translated into English and for which the corresponding italicized section in English seems to be a substitute):

> . . . I once thought that I would not live to be past thirty. That age is now long past. Gone, too, is the time when I craved my body's destruction. I recall someone saying at that time, "Let him holler; he'll soon wear himself out." Now what I hear is, "Let him be. It won't be long before he dies anyway." *I have lost my voice. Were I able to sing, would anyone hear this mute's soliloquy.* (*The Mute's Soliloquy*, 342)

> Dulu menduga umur takkan melewati tiga puluh. Yang tiga puluh terlewati. Terlewati juga detik-detik waktu tubuh yang sebatang ini diinginkan kehancurannya. Dulu kudengar orang berkata: Biar dia mencak-mencak kedungsangan, biar dia mati lelah sendiri. Yang sekarang terdengar lain: Biar saja, dia sudah tua, sebantar lagi akan

> mati sendiri. *Lupa orang bahwa setiap hidup ditutup oleh mati. Tak perduli serangga, margasatwa maupun manusia. . . .* [People forget that every life leads to death. Doesn't matter whether it's an insect, a wild animal, or perhaps humankind.] (199)

There is a striking contrast between the two sentences that give the title to the English translation and the two sentences they substitute from the Indonesian edition. To simplify, the one eloquently captures the defiant resilience of Pramoedya's dissident voice as an outspoken voice for human rights; the other dwells on the conditions of "bare life" in the camps.

My point is that the problem of translation that brings together Putu Oka Sukanta's "song of the starling" and Pramoedya's "mute's soliloquy" as two sides of the same "anthropological machine" compels us to attend to language, to a problem of language, that is neither animal nor man nor yet the chiastic passage between animal and man that leads Agamben to the formulation that sees the humanization of the animal and the animalization of the human as two sides of the same machine. In neither Pramoedya's "song" nor in Putu Oka Sukanta's "song" is there quite the same presupposition of language that Agamben traces in his analysis of the function of language in constructing anthropological assumptions (and the anthropological machine):

> In reality, the passage from animal to man, despite the emphasis placed on comparative anatomy and paleontological findings, was produced by subtracting an element that had nothing to do with either one, and that instead was presupposed as the identifying characteristic of the human language. In identifying himself with language, the speaking man places his own muteness outside of himself, as already and not yet human. (*The Open*, 34–35)

To some extent, both Pramoedya and Putu Oka might be seen to place their "own muteness" outside of themselves: Putu Oka, in figuring that "muteness" in the "song of the starling," and Pramoedya in repeatedly emphasizing the animal condition of man (at the end of the section quoted above, he offers a bitterly ironic version of the first line of Putu Oka's poem—"I write down the traces in a diary . . ."—"Such a pity it is that the deer and wild boar of this island will never be able to understand these notes of mine, even if they could be freed from their illiteracy" [343]). It is, however, the muteness inherent in language that is most enigmatic in each writer's "song"—and what contributes most to the power of their testimonial voices.

Muteness, in this sense, is the linguistic presupposition of the technology of starling song as we have traced it through all of these readings. First, as a matter of interspecies communication—as the demand for "listen"ing to the way the starling sounds and to the way the starling reproduces its surrounding sounds. Second, as a matter of language itself, the property (or faculty) of neither human nor animal but the singular and problematic passage of mimicry, mimesis, and translation across and between species. Third, and finally, as a matter of what lies outside, or to the side, of language—the paralinguistic register of whatever it is that starling song captures in the muteness of its technology.

Attentive to the disturbing dislocation of an experience of imprisonment, Putu Oka Sukanta's call to "listen" to the "song of the starling" is all the more insistent on a global positioning of the "starling song" the more particularly the word "jalak" ("starling") is inscribed within the endemic classification of the "jalak Bali" (Bali mynah). It is in this sense that Putu Oka Sukanta's "song of a starling" might be seen to replicate the technologies of biopower inherent in the technology of starling song: a technology that crosses over species, that defines the most distinctive faculties of human language, and that inscribes paralinguistic effects into the archive of the living or the death of species, language, and ecosystem.

NOTES

1. "Die Natur erzeugt Ähnlichkeiten. Man braucht nur an die Mimikry zu denken. Die höchste Fähigkeit im Produzieren von Ähnlichkeiten aber hat der Mensch. Die Gabe Ähnlichkeit zu sehen, die er besitzt, ist nichts als ein Rudiment des ehemals gewaltigen Zwanges, ähnlich zu werden und sich zu verhalten" (*Sprache und Geschichte*, 91).

2. See, in particular: "The formation of different languages and of distinct species, and the proofs that both have been developed through a gradual process, are curiously parallel. . . . Languages, like organic beings, can be classed in groups under groups; and they can be classed either naturally according to descent, or artificially by other characters. Dominant languages and dialects spread widely, and lead to the gradual extinction of other tongues. A language, like a species, when once extinct, never, as Sir C. Lyell remarks, reappears" (Darwin, *The Descent of Man*, 192–193).

3. "We've known for centuries that we possess five senses: sight, touch, hearing, taste and smell; but in reality there are several others including heat, cold, gravity, pain and acceleration. What's more, each of the five senses is actually a mixture of different sub-senses. Vision, for example, includes an appreciation of brightness, colour, texture and motion" (Birkhead, *Bird Sense*, xii).

4. See the *OED* definitions specific to starlings: "2a) a flock (of starlings); *spec.* (in later use) a large gathering of starlings creating intricate patterns in flight"; and "2b) the noise made by a flock of birds, esp. starlings."

5. For recent scientific studies and debates about starlings and what they show about the "language faculty" see, in particular, Hauser, Chomsky, and Fitch, "The Faculty of Language; Corballis, "Recursion, Language, and Starlings"; Corballis, "The Uniqueness of Human Recursive Thinking"; Jackenoff and Pinker, "The Nature of the Language Faculty"; Gentner et al., "Recursive Syntactic Pattern Learning by Songbirds"; Fitch, "On the Biology and Evolution of Music"; Anderson, "The Logical Structure of Linguistic Theory."

6. See, especially, Agamben and Derrida.

7. "Despite being seemingly disadvantaged relative to nightingales by a supposed lack of creativity, medieval parrots have the capacity to sing as well as talk, whereas nightingales can only sing, they cannot speak" (22).

8. Marcabru's starling is the very first example in Sarah Kay's book: "Around the middle of the century, the troubadour Marcabru wrote a brace of parodic love songs in which a foolish lover sends a starling—not quite a parrot, but the point is the same—to deliver his message to a tart who then turns him down" (1).

9. Cf. Heidegger's arguments about technology, Foucault's arguments about biopolitics, and perhaps especially Agamben's discussion of the "anthropological machine": "*Homo sapiens* . . . is neither a clearly defined species nor a substance; it is, rather, a machine or device for producing the recognition of the human" (*The Open*, chap. 7, "Taxonomies").

10. See Hollander, "The Sound of the Music of Music and Sound"; and his reference to Bloom, *Wallace Stevens*, 91.

11. Cf. also Bhabha, "Of Mimicry and Man," 91: "The 'unthought' across which colonial man is articulated is that process of classificatory confusion that I have described as the metonymy of the substitutive chain of ethical and cultural discourse. This results in the *splitting* of colonial discourse so that two attitudes towards external reality persist: one takes reality into consideration while the other disavows it and replaces it by a product of desire that repeats, rearticulates 'reality' as mimicry."

12. For an excellent, thoughtful, yet humorous account of Schieffelin's misguided effort to "bring the soothing, civilizing music of Shakespeare to New York's huddling masses" (174), see Mitchell, "The Bard's Bard; or, The Slings and Arrows of Avicultural Hegemony."

13. Yet with a slight typographical change, one might reread this as re-marking the starling trope "*in a starling transformation, the snowstorm becomes* . . ."

14. See the recent documentary films by Joshua Oppenheimer (*The Act of Killing* and *The Look of Silence*); Zurbuchen, *Beginning to Remember*; and the Buru novels of Pramoedya Ananta Toer.

15. "Insofar as the production of man through the opposition man/animal, human/inhuman, is at stake here, the machine necessarily functions by means of an exclusion (which is also always already a capturing) and an inclusion (which is also always already an exclusion) . . ." (*The Open*, 37).

WORKS CITED

Agamben, Giorgio. *L'aperto: l'uomo e l'animale*. Torino: Bollati Boringhieri, 2002.

———. *Homo Sacer: Sovereign Power and Bare Life*. Trans. Daniel Heller-Roazen. Stanford, Calif.: Stanford University Press, 1998.

———. *The Open: Man and Animal*. Trans. Kevin Attell. Stanford, Calif.: Stanford University Press, 2004.

Anderson, Stephen R. "The Logical Structure of Linguistic Theory." *Language* 84, no. 4 (December 2008): 795–814.

Aristotle. *On the Art of Poetry*. Trans. S. H. Butcher. Indianapolis, Ind.: Bobbs-Merrill, 1981.

Benjamin, Walter. *Reflections: Essays, Aphorisms, Autobiographical Writings*. Trans. Edmund Jephcott. New York: Schocken, 1978.

———. *Sprache und Geschichte: Philosophische Essays*. Stuttgart: Reclam, 1992.

Bhabha, Homi. "Of Mimicry and Man." In *The Location of Culture*. London: Routledge, 1994.

Birkhead, Tim. *Bird Sense: What It's Like to Be a Bird*. New York: Bloomsbury, 2012.

Bloom, Harold. *Wallace Stevens: The Poems of Our Climate*. Ithaca, N.Y.: Cornell University Press, 1977.

Corballis, Michael C. "Recursion, Language, and Starlings." *Cognitive Science* 31:697–704.

———. "The Uniqueness of Human Recursive Thinking." *American Scientist* 95, no. 3 (May/June 2007): 240–248.

Darwin, Charles. *The Descent of Man*. 2nd ed. 1874.

Deleuze, Gilles, and Félix Guattari. *A Thousand Plateaus: Capitalism and Schizophrenia*. Trans. Brian Massumi. Minneapolis: University of Minnesota Press, 1987.

Derrida, Jacques. *The Beast and the Sovereign*. 2 vols. Trans. Geoffrey Bennington. Chicago: University of Chicago Press, 2011.

Fitch, W. Tecumseh. "On the Biology and Evolution of Music." *Music Perception: An Interdisciplinary Journal* 24, no. 1 (September 2006): 85–88.

Gentner, T. Q., K. M. Fenn, D. Margoliash, and H. C. Nusbaum. "Recursive Syntactic Pattern Learning by Songbirds." *Nature* 440 (2006): 1204–1207.

GoGwilt, Chris. "The Voice of Pramoedya Ananta Toer: Passages, Interviews, and Reflections from *The Mute's Soliloquy* and Pramoedya's North American Tour." *Cultural Critique* 55 (Fall 2003): 217–246.

Graham, Jorie. *The Dream of the Unified Field: Selected Poems, 1974–1994*. New York: Ecco, 2002.

———. *Materialism*. New York: Ecco, 1993.

Hauser, Marc D., Noam Chomsky, and W. Tecumseh Fitch. "The Faculty of Language: What Is It, Who Has It, and How Did It Evolve?" *Science* 298 (2002): 1569–1579.

Hollander, John. "The Sound of the Music of Music and Sound." In *Wallace Stevens: A Celebration*. Princeton, N.J.: Princeton University Press, 1980.

Jackenoff, Ray, and Steven Pinker. "The Nature of the Language Faculty and Its Implications for Evolution and Language." *Cognition* 97 (2005): 211–225.

Kay, Sarah. *Parrots and Nightingales: Troubadour Quotations and the Development of European Poetry*. Philadelphia: University of Pennsylvania Press, 2013.

Mitchell, Charles. "The Bard's Bard; or, The Slings and Arrows of Avicultural Hegemony." In *Trash Animals: How We Live With Nature's Filthy, Feral, Invasive, and Unwanted Species*, ed. Kelsi Nagy and Phillip David Johnson II. Minneapolis: University of Minnesota Press, 2013.

Pramoedya Ananta Toer. *The Mute's Soliloquy*. New York: Hyperion, 1999.

———. *Nyanyi Sunyi Seorang Bisu: catatan-catatan dari Buru*. 2 vols. Jakarta: Hasta Mitra, 1995, 1997.

Putu Oka Sukanta. *Tembang Jalak Bali: Puisi Tertindas / The Song of the Starling: Poetry of Oppression*. Trans. Keith Foulcher. Yogyakarta: Pustaka Pelajar, 2000.

Stevens, Wallace. *The Palm at the End of the Mind: Selected Poems*. Ed. Holly Stevens. New York: Vintage, 1972.

Tiffany, Daniel. *Infidel Poetics: Riddles, Nightlife, Substance*. Chicago: University of Chicago Press, 2009.

Wallace, Alfred Russel. *Contributions to the Theory of Natural Selection*. 2nd ed. New York: Macmillan, 1871.

Zurbuchen, Mary, ed. *Beginning to Remember: The Past in the Indonesian Present*. Singapore: Singapore University Press, 2005.

CHAPTER 10

Yogini and Mynah Bird: On the Poetics and Politics of Transspecies Meditation

Madeleine Brainerd and Kaori Kitao

whose	are	if
meditative	we	this
gaze	reading	yogini
sees	the	is
the	depicted	as
bird	thing	tall
as	as	as
sufi	a	an
letter	script	*alif*

Meditating with a Mynah

MADELEINE BRAINERD

What follows is based on a recording of Madeleine Brainerd's remarks on the painting Yogini and Mynah Bird *by the so-called Dublin Painter (Chester Beatty Library, Dublin) on November 6, 2015, and on earlier meditations on the same painting, including a visit to the Metropolitan Museum's exhibit* Sultans of Deccan India, 1500–1700, *and in dialogue with Christopher GoGwilt and Kaori Kitao.*

To describe the painting, first. A mynah perches on the raised hand of a woman who stands amid flowering plants, wearing the red robe of an ascetic. She may be a mortal woman devoted to the Hindu god Shiva, or she may be a yogini (a tantric goddess or demigoddess). Her hair, top-knotted in the style of Shaivite yoga practitioners, reiterates the deep browns of the bird's feathers. Blossoms nearby reflect her clothes' vibrant colors. But her skin's blue-grays are echoed by the hues of the rocky middle and background, where a palace rises on a distant hill. Birds appear here and there, in pairs except for the mynah, who is paired only with the woman or

Yogini with a Mynah Bird. By the Dublin painter. Bijapur, early seventeenth c. Ink, opaque watercolor, and gold on paper, folio: 17 3/4×12 5/8 inches (44×32 cm). Trustees of Chester Beatty Library, Dublin (in 11A.31). Reproduced with kind permission of the Chester Beatty Library.

yogini. Art historians have debated the seemingly human figure's status as either a female ascetic or an otherworldly yogini, but there has been almost no discussion of the central figure as transspecies, a mynah together with the yogini or woman.

The painting was finished around 1605 in Bijapur, India, for a Persian court that paid respectful attention to Sufis as well as to yoginis. But let's take a step back for a moment. Persian and Sanskrit aesthetics of Bijapur were enmeshed in both the Islamic Sufi traditions of Deccan India and the Shaivite tantric and yogini traditions. So, there were at least these four possible overlapping protocols for the reception of the painting: (1) classical Sanskrit aesthetics, studied by the elite of Islamicate India; (2) the Sufi aesthetics and literary genre of romance (in which yogis, yoginis, and various birds figure); added to these, (3) the cultic texts and practices called tantra and yoga; and then (4) the Persian court's culture: Muslim, aristocratic, cosmopolitan, military, and somewhat more secular. I think all four of these, in one way or another, would have been active in the reception of the painting. Or better—we could call them (following Sheldon Pollock) transactive, since these traditions did not form isolated sectors so much as, to differing degrees, mutually influencing, intertwining, and merging currents—albeit with one or more current dominating others in various times and places.

Since the yogini or ascetic woman stands in the foreground, with the palace in the far distance, another question the painting raises right away is: What is her and the mynah's relation to the life of the palace? She wears those outrageously palatial clothes. Debra Diamond calls them "princess garb" and points out that both female human ascetics and otherworldly yoginis were depicted wearing it. Her red tunic is masculine in style, signifying a female ascetic (Navina Najat Haidar makes this point)—but a yogini may also wear that tunic. Her body's elongation indicates asceticism—she's lean from fasting. Her skin is gray because she is covered in ashes, presumably from the charnel grounds. Wearing ashes became a signature of tantric Shaivite practice—it was part of the rituals for gaining earthly powers or for rising above the cycles of life and death. Let's come back later to the questions of what might lie between her and the palace and whether she's human or otherworldly. I'll call her a yogini for now.

What interests me most is the relationship, in the center of the painting, between her and the mynah. In paintings influenced by the *citrasūtras* (Sanskrit texts on painting and sculpture), body position and gaze are significant. The yogini supports the bird in her hand with her index finger. But then that finger points downward, and the bird is holding itself upward, pointing its beak up toward the woman. The mynah tilts its head to the side and upward, I think, to the same degree that the yogini is tilting her head down toward the bird. One could dispute angles. But that's my sense of it. It's tilting up just as she's tilting down. They appear to be related to each other in this refined and balanced way.

The yogini's eyes are gazing up and perhaps to the right. Her upward gaze is called the meditative or contemplative gaze (there's a painterly convention for this). As for the mynah's eyes, I am pretty sure the physiology of the mynah prevents the bird from accomplishing that same gaze. I can't be sure—each time seeing the painting I couldn't be sure I was seeing the eye of the mynah. Here is what I think is going on. The mynah tilts its head as a bird does to see upward, mirroring both the yogini's head tilt and the yogini's gaze.

The *Samarāṅgaṇa Sūtradhāra* (an eleventh-century Sanskrit text) calls the contemplative gaze the *rasadr̥ṣṭi-yoginī*. (*Rasa-dr̥ṣṭi* is a glance signaling a work of art's *rasa*, its mood, "juice," or aesthetic experience.) This meditative glance is not just an aesthetic signpost. It even has an analogue today in psychiatric neurobiology. Bessel van der Kolk calls it the interoceptive gaze. During recovery from severe psychological trauma, this gaze signals a turning point. According to him, in order for recovery to proceed, the brain's limbic system (which has been altered by the trauma) must be accessed via the processes of the medial prefrontal cortex, which has "gone offline," as he puts it—and this beneficial access has taken place when the interoceptive, or meditative, gaze appears. Briefly put—among its many current uses, meditation helps rewire the limbic system, aiding in the recovery from trauma. In the painting *Yogini and Mynah Bird* we see this interoceptive or meditative gaze.[1] In the mutual head tilting, and in the parallel lines of the yogini's chin and the bird's beak, we might see the mynah and the yogini reflecting each other's attitudes.

Mirroring, between the bird and the meditative yogini, brings up the subject of *darshan*, relevant to all four of the major customs of reception. *Darshan* is a classical Sanskrit term from at least the second century CE. It comes from Odissi dance, a dance form that started in eastern India but moved west and south: It was common in Shaivite temples *and* occurred in the Mughal and Deccan courts by the sixteenth century. (Actually, *darshan* probably predates the dance, which brought it to prominence.) Odissi dance has an aesthetic that is more or less replicated in other Sanskrit art forms, but it started the ball rolling. Isabella Nardi points out that the *Viṣṇudharmottara Purāṇa* (an influential Sanskrit text with chapters on painting, circa perhaps 450–650 CE) advises painters to know the science of dance.

Darshan involves relationships, between dancer and audience and between dancer and divinity. The dancer's head and gaze direct the audience's attention. When that gaze is inward, the audience doesn't just recognize a signified inwardness but mirrors it: The audience feels what the performer feels. Likewise if the dancer performs a god, say Krishna, he also instantiates the god. So what gets set up is a mirroring between audience and performer and between performer and deity, through the gaze (*darshan*, like *dṛṣṭi*, comes from the root for "see"). *Darshan* is a reciprocal gazing. That is one way to define it. But it is a gazing that is metaphorical too. What gets mirrored is an accumulation of emotional and mental states that become an ambient state of feeling (*rasa*), prompted by the events of the drama and transmitted by the dancer. So, for example (following Uttara Asha Coorlawala's discussion of Odissi dance and *darshan*), the audience feels what Krishna/the actor feels, Krishna/the actor feels the audience feeling Krishna, and that intensifies what Krishna/the actor is doing, which in turn affects the audience, with increasing reciprocal intensity. The aim of the Odissi version is complete union with the god on the part of the actor and the audience—*ānanda* is the word for the bliss of it—or *samādhi*, the union—and that can also be a goal for Shaivite yogic meditation. Here's how one might see the bird and the meditating yogini tilting their heads: a suggestion of darshanic mirroring, a mutually meditative moment. That's one interpretive option.

I don't mean to say that the mynah is depicted as meditating. Nor that she isn't. (I say "she" because mynahs were usually imagined as female in Persian and Sanskrit literature.) Rather, that her communion with the yogini may involve a mutuality, an interchange of affects amenable to the yogini's meditation. The mynah's open beak signals difference as well as the reflected likeness between bird and yogini.

Some of the Hindu traditions that accepted the transmigration of souls allowed that nonhuman species could become fully enlightened, achieving *samādhi*. (One can visit the gravesite of a "fully realized" crow, among those of other "emancipated" nonhumans, at an Advaita Vedanta ashram in Tiruvannamalai today.) Sufis, however, rejected theories of transmigration; birds in Persian and Hindavī Sufi narrative poems reflect their god by singing the praises of Allah or through demonstrations of wisdom. In the *Padmāvat*, a Sufi romance, the parrot Hīrāman, who is the royal tutor, has incomparable wisdom among Brahmins. That's a parrot, not a mynah, but parrot and mynah wisdom were closely associated. In the Persian *Ṭūṭī-Nāma*, a mynah risks and loses her life to speak wisely.

The Sufi context matters, partly because Ibrahim ʻAdil Shah II (r. 1580–1627), who likely commissioned the painting, took an interest in Sufis—this was common in the Mughal and Deccan courts. Sufi shaikhs were powerful religious leaders whom the secular powers needed both to court and to support. Sufi romances were read in the Persian courts and their poetics defended in Islamic courts of law: In transcultural India, they participated in the mix of elite aesthetics. That doesn't make *Yogini and Mynah Bird* a Sufi painting, but it does raise the question of how Sufism may have contributed to its interpretations. A glance at the sultanate will lead us back to the Sufis.

Sultans also found the tantric yoginis interesting. Ibrahim ʻAdil Shah's uncle, Sultan Ali ʻAdil Shah I, compiled a manuscript (the *Stars of the Sciences*), which included descriptions of 140 yoginis; as others have noted, he stated that he was propitiating yoginis because he wanted their military assistance. Tantric yoga sometimes aimed not so much to seek peaceful, blissful union with the divine as to cultivate worldly power, in cults of clan and personal empowerment. Supernatural abilities (*siddhis*) were attributed to yoginis. So sultans requested their help. The yoginis' association with Shaivism does not seem to have been a barrier to that request. The nephew, Ibrahim, was also reputed to be devoted to a Hindu goddess, Saraswati, the goddess of music and education, although he was undoubtedly Muslim. The sultans seem to have moved easily among a number of different cultural spheres.

The yoginis in the *Stars of the Sciences* were magical, portrayed as Persian *ruhaniya* (earth-spirits) who possessed special *siddhis*. Worship of yogini precursors had probably begun centuries earlier with local bird, animal, and tree goddesses and demonesses (David Gordon White discusses this); if the painting evokes a deep connection between bird and yogini, that genealogy may be relevant. Nath yoga, which had inherited and adapted Kaula (tantric) yogini rites, was known to the Persian courts and to the Sufis.

But the woman in the painting, as we've seen, has some attributes of a human ascetic practitioner. If mortal, however, she's dressed as masculine and feminine, ascetic and princess; if immortal, as all those—and as mortal. Perhaps she also appeared as a sublimely beautiful woman, to whoever added, about a hundred years after the painting's completion, the Persian verses that frame it. More on them shortly; first, the reflections Sufis saw in feminine beauty.

Sufi clerics defended their narrative-verse genre of romance in courts of Islamic law by elaborating a rhetoric of women's attractiveness. According to Aditya Behl's discussion of their poetics, the first time the hero sees the woman who will become his beloved—in a typical plot—her beauty gives him an inkling of the allure of his God. This inkling prompts for the reader or auditor and (allegorically) for the hero a memory of the past, when his soul was united with the divinity. For the group of Sufis (mainly Chishti and Shattari) writing romances, the divine here could be described as a cosmic love pervading the universe. Unlike Shaivites, Sufis rejected divine incarnation as well as spiritual reincarnation, but the memory of past unity has a present force that impels the hero and reader toward seeking it again. The divinity becomes present, for an educated Sufi reader, as mirrored love and desire in the reader's (and the hero's) heart, analogous to a reflection of light. As if his heart mirrored a lamp that shone from above, its rays refract though the woman's beauty into him.

Heroine and hero then become separated in the formulaic narrative. Here is one way the lines of verse inscribed in the borders of the painting relate to the woman in the painting. Sanskrit poets wrote what is called toe-to-head verse, describing a woman from toe to head. When these were adapted by Hindavī Sufi romance for depicting the beloved during the period of separation, they became head-to-foot descriptions. Head-to-foot poems present the woman's attractions in ways that reiterate their relationship to the hero and the reader as a mirror of cosmic love. Earlier Persian verse styles and attitudes like those in the painting's frame had become incorporated into romance's head-to-foot verses. While the lines in the frame offer an insight into a later reading, they also typify romance poems that Sufis at the time of the painting read with pedagogical, mystical purpose.

The Sufi poet Shaikh Kamal Khujandi (d. 1400) wrote the frame's couplets comparing a woman's perfect appearance to the letter *alif.* Shattari Sufis in the following centuries performed a version of alphabet mysticism. Each letter in the names of Allah was separately meditated on. Concentrating on a letter, the practitioner invited divine affection to be reflected within him. (The Sufi orders were all male.) Letters became cues for the transformative arrival of that heavenly affect. Alphabet mysticism like that prefigured by Kamal Khujandi's poem occurs in allegorized head-to-foot verse, reinforcing the possibility of a romance reading of the painting: The woman's appearance, likened to an *alif,* prompts a believer to let earthly passion become devout. Written earlier than the painting and added later, the poem bears witness to the continuity of Sufi practice from the fifteenth to the eighteenth century, employing a rhetoric of affection mirrored from above for responding to female beauty.

Sufi romance heroines may be read allegorically as merely reflective surfaces, although the texts sometimes present them as more than that. Likewise, to read the painting allegorically would be to connect the viewer's feelings with his deity via the woman as looking glass, bypassing both her and the mynah's innerness, but as some romances suggest feminine interiority, so too does the painting's meditative glance.

Tantric yoga's own alphabet mysticism contributed to Sufi theory and practice. Tantric texts had advised chanting Sanskrit syllables to empower the meditator, by aligning human energies with those of a great yogini such as the goddess Shakti, since at least the eleventh century.

But yoginis figure in Sufi romances only as minor supernatural agents, furthering events of the plot. Nor are they creatures of beauty—beauty is for heroines. The heroine, separated from her beloved, can find that she's been turned into a bird; in other tales she requests aid from magical beings—one of whom might be a yogini. The painting lends itself to a variety of narrative readings without granting any one of them. Is a heroine (as bird) beseeching a yogini to reunite the lovers? Or has a princess disguised herself as a yogini? (M. L. Nigam suggests this.) Is the bird her companion, glancing quizzically at its human, or is it her tutor?

Any of the three—a yogini, a female ascetic, or a mynah—could be an adept or holy teacher, depending on which Persian or Sanskrit narrative or religious contexts one emphasized. A Shaivite ascetic woman might meditate while a bird mirrored her *samādhi*.[2] But mynahs (and parrots) had the literary reputation of being tutors. The heroine of the *Padmāvat* relies on the wise parrot I mentioned earlier to teach her how to find her lover. Literary mynahs tend to be mates of parrots, as in the *Śukasaptati* (translated into Persian as *Ṭūṭī-Nāma* by a Shattari shaikh), in which the mynah is the first to give a wife good advice.

When a disciple gazes into the eyes of his teacher, according to Sufi pedagogy, the teacher's inward focus may be so powerful that it transforms the disciple, similarly to a yogic darshanic transference of innerness between guru and student. Disciple reflects teacher reflecting contemplative affect. I should mention that Sufism, especially Shattari Sufism, borrowed more from tantra (through the Naths) than from any other yogic branch. And yogic practice began to resemble some aspects of Sufism. There's a lot of mirroring going on.

Thinking of the four most obvious protocols for reading the painting—Sanskrit, Sufi, tantric, courtly—courtly interpretation tended to be more secular in approaching its pleasures: elegant woman with companion bird (this was an Indo-Persian pictorial motif), refinement of line and color, abundance of interpretive options. Sanskrit, Sufi, and courtly poetics valued suggestiveness, the generation of multiple meanings and shades of meaning. Sanskrit theories extended that maximal allusiveness to painting. Behl points out that as visitors from a variety of cultures passed through the sultans' centers of power, they might see or hear works of art that allowed suggestiveness, *vyañjakatva* or *dhvani*, to give rise to disparate interpretations relevant to their sundry backgrounds.

The fact that the mynah's beak is open (which you can't see very well in this reproduction) suggests that it may be the mynah who is singing or teaching, or—and this is equally a possibility—that the mynah is nipping at her earring while the yogini's or woman's meditation includes enjoyment of the bird's playfulness.

Such a courtly version brings us back to the palace. The cosmopolitanism of the sultan's court embraced, to varying degrees, Persian as well as Arabic, Turkish, Rajput, and other Indic, Himalayan, and Chinese aesthetics. I think the painting invites at least all four of the overlapping cultures of reception. At its dynamic center, a bird-and-woman's mimicry draws attention to the painting's own aesthetic mirroring—its Chinese flower blossoms, Shaivite hair style, Bijapur landscape, and other cultural borrowings. Shatha Almutawa writes about Deccan painting as collage. *Yogini and Mynah Bird* seems to me to make use of that collage aesthetic in optimizing its interpretive options.

A multiply mirroring, perhaps mutually meditative, dyadic figure mediates the painting's own recombinatory finesse. That foregrounded pair also upstages the distant palace. Still, the fact that yoginis' powers were sought by the sultan's uncle (and by other sultans and kings) for warfare brings to mind another side of meditation. Meditators offer their practice today as something needed by cultures that are traumatized, stressed, aggressive. And meditation does help reduce the "fight-or-flight" reactivity of a stressed populace; its usefulness for recovery from trauma even includes rendering compassion more available. But there are also historical instances of violence from cultures that meditate. The U.S. Army's use of yoga and meditation, for example, increases its soldiers' mental health and military efficiency. Certain yogis in India from the sixteenth into the eighteenth centuries (Shaivite and perhaps Vaishnavite) were mercenaries; they hired themselves out and also fought among one another for water rights. They would establish themselves on holy riverbanks, charge money for water, and then fight other territorial gangs of yogis.

Meditation, in a sense, is like deconstruction. You can use it for any purpose. For it to calm the limbic system, a particular phase of meditation has to happen: a nonverbal phase. If the verbal system (involving the frontal brain) is engaged, that's going to shut off limbic access. Neurological studies show that to facilitate trauma reduction there has to be that nonverbal moment. (This doesn't mean that when a meditative state is established, you can't be aware of something else going on, hear whatever a mynah is saying, or grasp what your teacher is advising. You can set it up and then become aware of what else is happening—consider mindfulness meditation, with its simultaneous awareness of "inner" and "outer" experience.)

When you meditate, you activate mirror neurons. Many mammals have mirror neurons—and birds have them, or similarly functioning cells, too. If a baby smiles because the person facing it is smiling, mirror neurons have been activated. They affect the motor system and other mind-body systems. Antelopes galloping together, wheeling about as a unit—that's mirror neurons working. Or humans synchronizing on a dance floor. And mirror neurons have to be activated for empathy to take place. Meditation's activation of mirror neurons is now being considered hypothetically (by the neuropsychiatrist Daniel Siegel) as the reason that meditation deepens compassion. Studies have been done already (as skeptical as we may be) indicating that people score higher on compassion after meditation than before. And the more deeply they meditate, the higher their compassion score. My sense of it, given the different ways you can use meditation, is that even though it *deepens* compassion, meditation might not *extend the range* of compassion. A soldier who practiced it might then harbor greater care for whoever was already an object of his or her care but possibly not for those designated as enemies. Perhaps a verbal intervention, or some other kind of experience, is required for extending the range.

So. The yogini/woman relates to the mynah and the mynah to her. They reflect each other, but they also stand before this palace. Yogini meditation sometimes provided initiates with martial prowess. The yogini may be a mercenary of sorts, willing to aid the palatial military, so long as she gets properly propitiated. As Rajput, Mughal, Turkish, and Persian forces fought one another across the decades, the conflicts a yogini might be persuaded to contribute to were destroying regional habitats of mynahs and of the peasants who trapped mynahs to sell to the nobility. Suffering, devastation—are not to be seen in the painting. Unless, of course, we consider them suggested by the ashes from the charnel grounds.

And if she is an ascetic follower of Shiva? She would be subject to Persian power. We can sense a fraction of the complexities of assimilation, estrangement, and even respect that that might entail, by noticing that her form has been exoticized.

But the yogini/woman and mynah look so tranquil. It's tempting to call their apparent serenity the *rasa* of the painting. This would be *śānta rasa*, as the *Viṣṇudharmottara Purāṇa* calls it. But assigning a *rasa* engages our own angles of refraction, materials of reflection. Mirroring by the mind (which, in seventeenth-century India, can include the emotions) is a learned behavior. Sufis trained initiates in poetry reading; Muslim and Hindu aristocrats refined their aesthetic responses. A lineage of gurus taught Naths to introject divine powers. The mirroring gazes they practiced—honing a specific openness to the seen and its associated unseen, in each case—deliberately induced transmission and transformation of the affects. Maybe it's true, as some neurologists say, that affective states govern personal and social thoughts and actions. At any rate, the model of influencing the interaffectivity they lived in—fields of feelings that might cross both species and cultural boundaries—by modulating felt responses to other beings, refracting and reflecting, obtained among tantric ascetics, Sufis, and throughout elite pedagogy.

Meditation with a mynah can take many forms.

Further Meditations

Kaori Kitao

Friends called my attention to this remarkable painting *Yogini with Mynah Bird* (Freer/Sackler, 1603/1604, opaque watercolor and gold on paper, India Karnataka, Islamic court of Bijapur), and it prompted me to gaze at it and scrutinize it.[3]

The arbitrary scale in the landscape, like the overscaled plants in the middle ground, is conventional enough in a non-European work like this one; the oddity is apparent only when viewed with the perspectival perception ingrained in the Western tradition. The detail that I find befuddling is the bird, perched on the woman's right hand, which turns its head completely upside down, anatomically a well-nigh impossible position. With its beak close to the mouth of the woman, the artist, we can be certain, designed it to draw the viewer's attention to the relationship between the yogini and the bird.

Reading a picture is a risky business. Since the mynah, like the parrot, is a bird that "talks," one may be tempted to understand that the bird is conversing with the yogini, or, since the woman's mouth is closed, she may be only listening to what the bird is saying. One interpretation mistakenly speaks of "the lady whispering to her talking bird."[4] This writer, moreover, doubts the lady is a yogini but believes her to be possibly "Balqis, the Queen of Sheba." Another writer reads the yogini's gaze differently: "The mynah, perched on her arm was playfully speaking with her or maybe tugging at one of the pearls from her earrings but the yogini appeared to be in a trance, looking beyond the bird, beyond everything."[5] The yogini may *seem* to be whispering to the bird or only listening to it. She may *seem* to be gazing or contemplating, or she may be merely gazing without seeing anything. Then, the mynah may be talking or just listening, except that a mynah, by convention, is understood to talk. When we read a picture of a bird or any animal, we engage ourselves in an anthropomorphic projection.

Mention of anthropomorphism may at first seem to make light of the sense of deep communication between the yogini and the bird. But instilling a human character in any nonhuman creature, from mammals down to fishes, is an act of *compassion*, and it allows us to talk, say, to a fish, even though it may not respond. Compassion is one-way traffic, unlike communication that assumes correspondence, that is, answering the other. Between two individuals, on the other hand, there is never complete comprehension. One can never fully fathom, from the words spoken, what is in the mind of one's interlocutor. But understanding occurs without comprehension, as described in this Zen *mondō*: A Zen master, taking a walk with his disciple beside a pond on a beautiful day, said, "Look, how happy those fish are in the pond," and the disciple retorted, "Master, you are not a fish, so you cannot really know if they are happy or not." The Master then replied, "You are not me, so how do you know what I had in my mind when I said what I said?" Compassion enables us to empathize and trust what was meant by what was said, though ultimately we can never be sure.

To say that a mynah or a parrot talks means that it imitates sonically human speech. When we humans try to imitate the sounds made by birds and beasts, our effort is onomatopoeia, and it is a crude approximation: "meow," "oink," "tweet tweet." A mynah does not merely imitate spoken words but the pitch and tone, as well as the timbre, of the vocalized phrase it has been taught to repeat by its trainer. If we had this capacity to imitate, we could learn to speak a foreign language without any accent. A child learns its mother tongue by close imitation, much closer than an adult can trying to speak a foreign language—but still not as close as a mynah or parrot. When we think a mynah is talking when it is only repeating the sound, we are acting anthropomorphically and think by compassion that we are being understood. Seen in this light, we can consider that the yogini is communicating with her mynah in this image, and in so understanding we are in turn reading the picture compassionately. So, the bird, momentarily appearing supine but more likely upright, *seems* as though it is making a tremendous effort to communicate with his yogini, and therefore the yogini in turn *seems* as though she is contemplating without responding directly to the bird's beckon yet understanding it compassionately. This may ultimately be the meaning of the painting. Reading an image is quite like reading the mind of another.

NOTES

1. Juxtaposing a seventeenth-century cultural practice with twenty-first-century neurology admittedly has its risks. For instance, customary practices like meditation might not have engaged the same neural networks then as now. Also, as neuroscience evolves it may reject versions of a "limbic system." Still, I hope that considering present with past explanations can add a dimension of thought, rather than reducing one to the other.

2. The general question of whether a mynah in "real life" can meditate may seem preposterous, at first, but it is, in part, an empirical matter. Humans living with mynahs say that the birds occasionally exhibit a relaxed state of diffuse awareness. If while in that state they have a slowed heartbeat, appropriately activated neurons, and the other physiological signs of meditation, then they probably are in something like a meditative state. (Jennifer Ackerman reports that for now, differences between bird and human brain structures make parallel mapping impossible, but ornithologists have found that despite dissimilar neural patterns, birds and humans perform similar cognitive feats, so a more nuanced understanding of avian nerve systems is under way.)

More important is the principle of allowing parallels with human behavior to imply cognitive and affective similarity in other species—as the default position in our regard for nonhumans. Frans de Waal developed this principle, which renders anthropomorphism largely irrelevant in many cases of attributing experiences to nonhuman species. On the other hand, in the romances and frame tales, birds who give advice in human languages have been anthropomorphized. I'm grateful to Kaori Kitao for raising the question of anthropomorphism in the context of *Yogini and Mynah Bird.*

3. This section is taken from Kaori Kitao's blog: http://kaoriswhirlinggig.blogspot.co.uk (November 18, 2015).

4. http://www.indianart.ru/eng/of_the_deccan/7.php.

5. http://mariecameronstudio.com/tag/yogini-with-myna/.

WORKS CITED

Ackerman, Jennifer. *The Genius of Birds*. New York: Penguin, 2016.

Almutawa, Shatha. "A Seventeenth-Century Indian Parrot: Interpreting the Art of the Deccan." *Perspectives on History*, July 2015. http://historians.org/publications-and-directories/perspectives-on-history/summer-2015/a-17th-century-indian-parrot.

Behl, Aditya. *Love's Subtle Magic: An Indian Islamic Literary Tradition, 1379–1545*. New York: Oxford University Press, 2012.

Behl, Aditya, and Simon Weightman. *Madhumālatī: An Indian Sufi Romance*. Oxford: Oxford University Press, 2000.

Coorlawala, Uttara Asha. "Darshan and Abhinaya: An Alternative to the Male Gaze." *Dance Research Journal* 28, no. 1 (September 1997). https://www.researchgate.net/publication/274921862_Darshan_and_Abhinaya_An_Alternative_to_the_Male_Gaze.

Dave Mukerji, P. *The Citrasutra of the Visnudharmottara Purana*. Delhi: Indira Gandhi National Centre for the Arts and Motilal Banarsidass, 2001.

De Waal, Frans. *The Age of Empathy*. New York: Random House, 2009.

Diamond, Debra. "Occult Science and Bijapur's Yoginis." In *Indian Painting: Essays in Honor of B. N. Goswamy*, ed. Mahesh Sharma and Padma Kaimal. Ahmedabad: Mapin, 2013.

———. "Yoginis." In *Yoga: The Art of Transformation*, ed. Debra Diamond. Washington, D.C.: Smithsonian Institution, 2013.

Haidar, Navina Najat. "Yogini with a Mynah Bird." In *Sultans of Deccan India, 1500–1700: Opulence and Fantasy*, ed. Navina Najat Haidar and Marika Sardar. New York: Metropolitan Museum of Art, 2015.

Hansar, A. N. D., trans. *Shuka Saptati: Seventy Tales of the Parrot*. New Delhi: Harper Collins, 2000.

Nardi, Isabella. *The Theory of Citrasūtras in Indian Painting*. New York: Routledge, 2006.

Nigam, M. L. "The Yogini of the Deccani Miniatures." *Lalit Kala* 23 (1988).

Nujum al 'Ulum (The Stars of the Sciences) by Sultan Ali 'Adil Shah, 1570–1571. Manuscript at the Chester Beatty Library, Dublin.

Pollock, Sheldon. *The Language of the Gods in the World of Men: Sanskrit, Culture, and Power in Premodern India*. Berkeley: University of California, 2006.

Sastri, G., ed. *Samaranganasutradhara of King Bhojadeva*. Vol. 2. Gaekwad's Oriental Series 25. Baroda, 1925.

Schmitt, Richard, ed. *Die Cukasaptati, Textus Simplicior*. Leipzig: F. A. Brockhaus, 1893.

Shirreff, A. G. *Padmavati of Malik Muhammad Jaisi*. Bibliotheca Indica 267. Calcutta: Royal Asiatic Society of Bengal, 1944.

Siegel, Daniel. Lecture at Kripalu Center for Yoga and Health, Stockbridge, Mass. September 25, 2014.

Simsar, Muhammed A., trans. and ed. *The Cleveland Museum of Art's Ṭūṭī Nāma Tales of a Parrot by Nakhshabi, Ziya' U'd-Din*. Cleveland: Cleveland Museum of Art, 1978.

Van Der Kolk, Bessel. *The Body Keeps the Score: Brain, Mind, and Body in the Healing of Trauma*. New York: Viking, 2014.

———. Lecture at Kripalu Center for Yoga and Health, Stockbridge, Mass. April 5, 2014.

White, David Gordon. *Kiss of the Yoginī: "Tantric Sex" in Its South Asian Contexts.* Chicago: University of Chicago, 2003.

Plate 1. Page from *Tales of a Parrot (Tuti-nama): First night: Khujasta kills the pet myna who advises her not to be unfaithful to Maymun, her husband*, c. 1650. India, Mughal, Reign of Akbar, 16th century. Opaque watercolor, gold and ink on paper; page: 19.9 × 14.2 cm (7 13/16 × 5 5/8 inches). The Cleveland Museum of Art, Gift of Mrs. A. Dean Perry 1962.279.8.a. Reproduced with kind permission of the Cleveland Museum of Art.

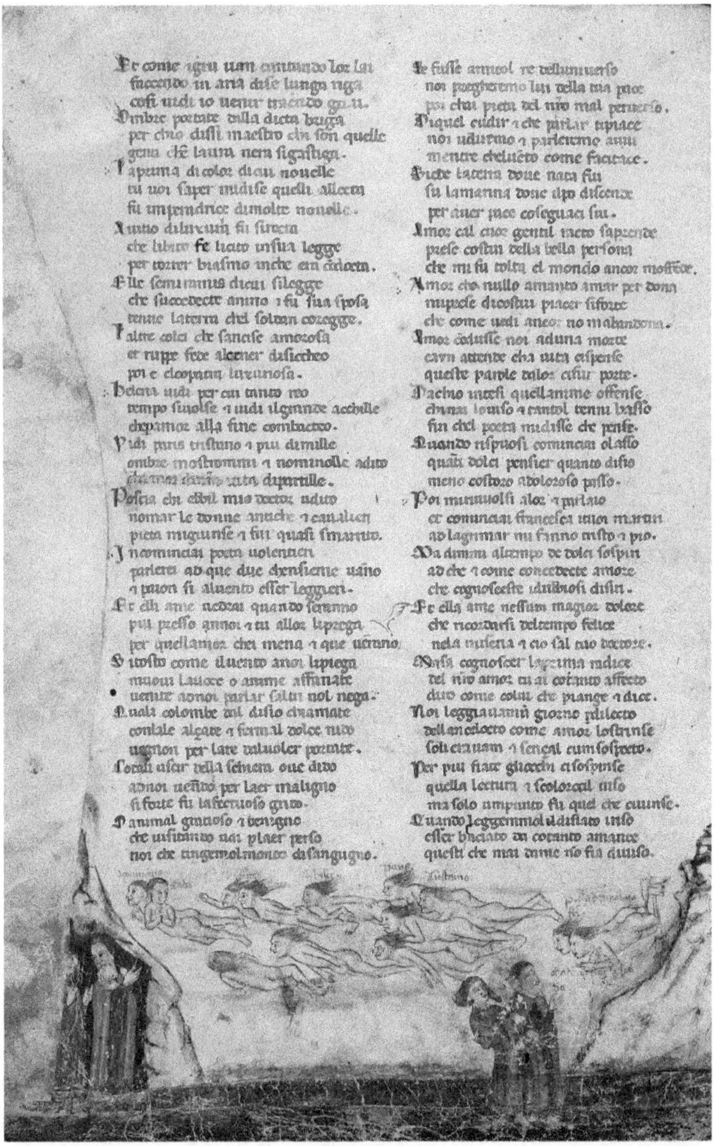

Plate 2. The crowd of souls in the second circle of Hell, figured as starlings or cranes, from fourteenth-century illuminated manuscript of Dante's *La Divina Commedia*: Bodleian Library. MS. Holkham Misc. 48, p. 8; page of illustrated MSS of Dante's *Inferno* (Canto V). The comparison of the "carnal sinners" to cranes appears at the top of the page: "E come i gru van cantando lor lai" ("And as the cranes go chanting their lays"). The stanzas just before that compare the souls to starlings: "E come li stornei ne portan l'ali / nel freddo tempo" ("And as their wings bear the starlings along"). See Dante Alighieri, *The Divine Comedy*, trans. Charles Singleton (London: Routledge & Kegan Paul, 1971), 50.

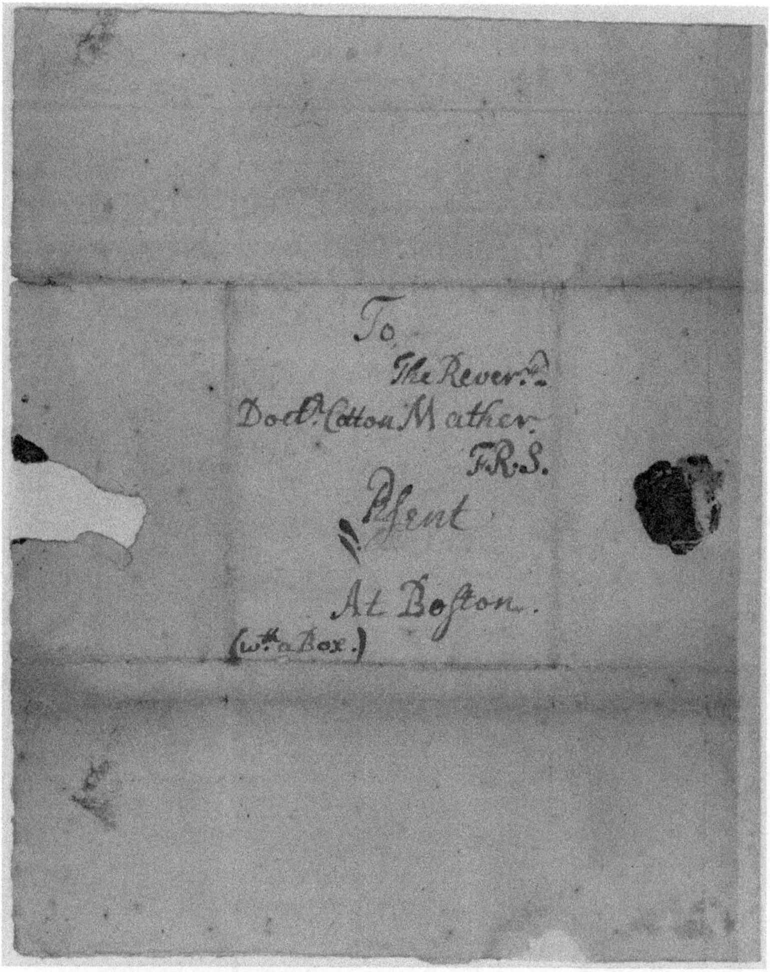

Plate 3. Envelope of John Winthrop's letter to Cotton Mather, "with a box" (evidence of a bird specimen). Benjamin Colman Papers, 1641–1806. John Winthrop (New London) to Cotton Mather, December 29, 1719. Mz. N-1013. Collection of the Massachusetts Historical Society.

Plate 4. Carel Fabritius, *The Goldfinch*. 1654. Mauritshuis, The Hague.

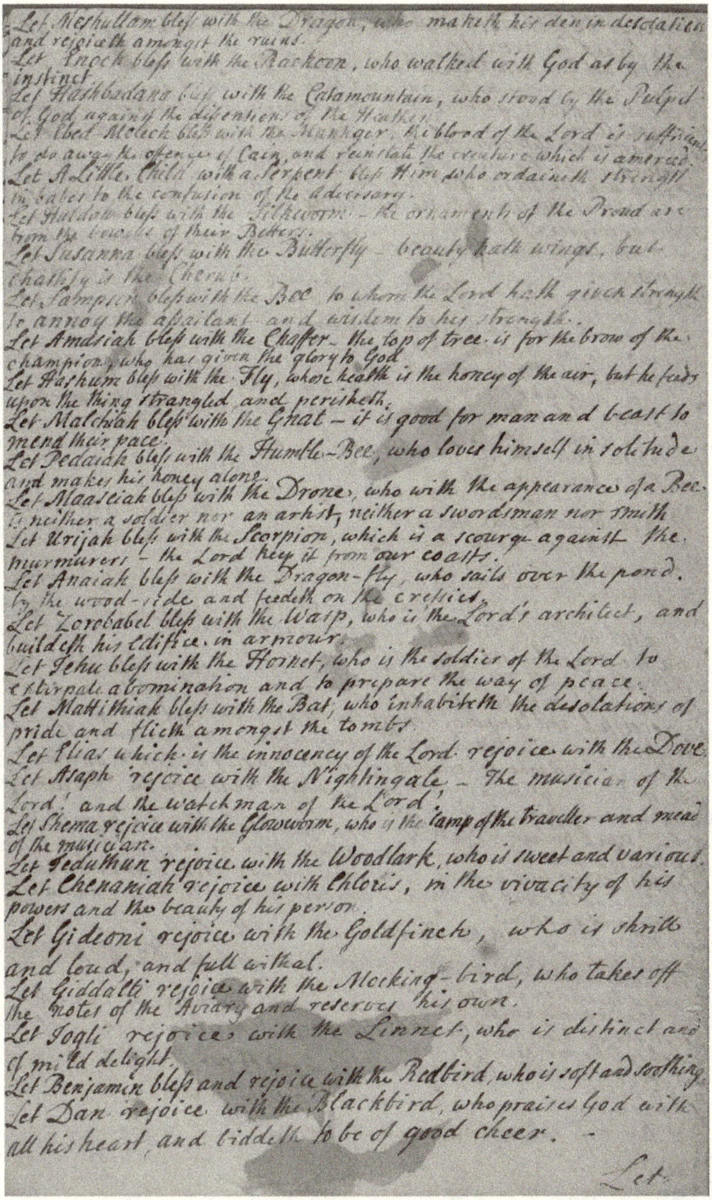
Let Meshullam bless with the Dragon, who maketh his den in desolation and rejoiceth amongst the ruins.
Let Enoch bless with the Rackoon, who walked with God as by the instinct.
Let Hashbadana bless with the Catamountain, who stood by the Pulpit of God against the dissensions of the Heathen.
Let Ebed-Melech bless with the Mantiger, the blood of the Lord is sufficient to do away the offence of Cain, and reinstate the creature which is amerced.
Let A Little Child with a Serpent bless Him who ordaineth strength in babes to the confusion of the Adversary.
Let Huldah bless with the Silkworm – the ornaments of the Proud are from the bowels of their Betters.
Let Susanna bless with the Butterfly – beauty hath wings, but chastity is the Cherub.
Let Sampson bless with the Bee to whom the Lord hath given strength to annoy the assailant and wisdom to his strength.
Let Amasiah bless with the Chaffer – the top of tree is for the brow of the champion, who has given the glory to God.
Let Hashum bless with the Fly, whose health is the honey of the air, but he feeds upon the thing strangled and perisheth.
Let Malchiah bless with the Gnat – it is good for man and beast to mend their pace.
Let Pedaiah bless with the Humble-Bee, who loves himself in solitude and makes his honey alone.
Let Maaseiah bless with the Drone, who with the appearance of a Bee is neither a soldier nor an artist, neither a swordsman nor smith.
Let Urijah bless with the Scorpion, which is a scourge against the murmurers – the Lord keep it from our coasts.
Let Anaiah bless with the Dragon-fly, who sails over the pond by the wood-side and feedeth on the cresses.
Let Zorobabel bless with the Wasp, who is the Lord's architect, and buildeth his edifice in armour.
Let Jehu bless with the Hornet, who is the soldier of the Lord to extirpate abomination and to prepare the way of peace.
Let Mattithiah bless with the Bat, who inhabiteth the desolations of pride and flieth amongst the tombs.
Let Elias which is the innocency of the Lord rejoice with the Dove.
Let Asaph rejoice with the Nightingale – The musician of the Lord! and the watchman of the Lord!
Let Shema rejoice with the Glowworm, who is the lamp of the traveller and mead of the musician.
Let Jeduthun rejoice with the Woodlark, who is sweet and various.
Let Chenaniah rejoice with Chloris, in the vivacity of his powers and the beauty of his person.
Let Gideoni rejoice with the Goldfinch, who is shrill and loud, and full withal.
Let Giddalti rejoice with the Mocking-bird, who takes off the notes of the Aviary and reserves his own.
Let Jogli rejoice with the Linnet, who is distinct and of mild delight.
Let Benjamin bless and rejoice with the Redbird, who is soft and soothing.
Let Dan rejoice with the Blackbird, who praises God with all his heart, and biddeth to be of good cheer. –
Let

Plate 5. Christopher Smart, from *Jubilate Agno*. 1759–1763. Houghton Library, Harvard, MS Eng 719. The manuscript consists of over 2,100 lines, some beginning with "Let," others beginning "For." Although scholars continue to discuss how the poem as a whole should be organized, Smart's intention was to organize the "Let" and "For" lines in corresponding columns, to be read antiphonally. The current page is from Fragment A. "Let" seq. 4. and features (among other birds) the "Mocking-Bird."

Plate 6. Robinson Crusoe: engraved map of the island by John Clark and John Pine (1719). Daniel Defoe, Serious reflections during the life and surprising adventures of Robinson Crusoe: with his Vision of the angelick world / written by himself. Reproduced by kind permission of the University of Pennsylvania Kislak Center for Special Collections, Rare Books and Manuscripts.

[38]

I have borne this poor ſtarling as the creſt to my arms.—Thus :

——And let the heralds officers twiſt his neck about if they dare.

THE

Plate 7. Laurence Sterne, *A Sentimental Journey* (London: T. Becket & P. A. de Hondt, 1768), 38. Image of heraldic starling, Parson Yorick's coat of arms. Reproduced by kind permission of the Syndics of Cambridge University Library.

Plate 8. Franz Schubert, "Ungeduld" ("Impatience") autograph music score from *Die Schöne Müllerin* (*The Fair Miller Maid*). Reproduced with permission of the Wienbibliothek im Rathaus, Musiksammlung. The second verse reads, in part: "I'd like to train a young starling / To speak the words clearly and distinctly / So that it would speak with the sound of my voice . . ."

Plate 9. Serinette, bird organ. Paris. C. 1770. © Victoria and Albert Museum, London. Made to teach caged birds to sing different tunes, the bird organ is operated by a handle that pumps the bellows in a wind-box and rotates a cylinder controlling the air supply to organ pipes. Different kinds of bird organ were made in France, such as a *merline*, which copied blackbirds, and a *turlutaine*, which copied curlews. This is a *serinette*, made to copy the sound of finches. The case for this bird organ was made by Leonard Boudin (1737–1807).

Plate 10. John James Audubon, *Common Mocking Bird*. From John James Audubon, *Birds of America* (1827–1838). From the New York Public Library Digital Images Collection.

Plate 11. John James Audubon, *Red-winged Starling*. From John James Audubon, *Birds of America* (1827–1838). "If the name of Starling has been given to this well-known species, with the view of assimilating it to the European bird of that name, it can only have been on account of the numbers of individuals that associate together, for in every other respect it is as distinct from the true Starlings as a Common Crow" (Volume 1, 348). From the New York Public Library Digital Images Collection.

Plate 12. "Floreana Mockingbird," on 2009 British 81 p postage stamp, celebrating the 200th anniversary of Charles Darwin's birth. © Stamp Design Royal Mail Group Ltd (2009).

AFTERWORD

A "Starling" Manifesto for Mocking Bird Technologies

Christopher GoGwilt

scroll	the	see
down	checker	a
words	set	starling
in	of	in
criss-cross	*jalak*	a
flight	sounds	tree
of	mimic	and
figure	making	think
eights	others	again

There is a natural urge to classify critical essays such as those collected in this volume according to this or that direction in critical studies. This collection is designed to resist such "pigeonholing" and to cross over—deliberately, theoretically, and also eclectically—a range of disciplinary and interdisciplinary categories. The resistance to classification itself, of course, might be interpreted as a kind of theoretical manifesto. There is no resistance to theory that is not itself making some theoretical claim. The critique of classifications enacts its own revenge in the very form of its own critique. Maybe the best way to underscore this volume's idiosyncratic idiom of critical engagement with mocking bird technologies is to give it a mocking bird manifesto of its own. That is what I offer here, in this afterword, in explaining the specific technology of "starling" form used in the mock précis for each essay and in discussing its relation to the current collection of essays.

"Starling" form, as I have developed it over the past few years, is a technology of writing more like a set form of lyric poetry (sonnet, haiku, sestina, or pantun) than the jargon prose of theory. Unlike the recognized forms of poetry, however, starling form is entirely personal (I am almost the only one who uses this form),[1] and it is highly unlikely to become established as a recognized form of poetry. It is a form that I propose (manifesto-like) exists somewhere between the ambitions of poetry and the ambitions of theory. Its technology of the word emulates the technology of starling song (as I discussed in Chapter 9), mimicking both the behavioral patterns of actual starlings and the problem of classification posed by the bird and the word. In this respect, the form poses and probes many of the strands of argument that run through all the essays in this collection: for example, the question of citation and quotation central to literary criticism and also lyric form, the problem of a distinction between prose and poetry, the difference between overlapping but distinct literary traditions, the rhetoric of affect and address, the interrelation between the idioms of critical theory and creative practice, and the relation between classifications of birds and words.

During the process of co-organizing the ACLA seminar from which this volume has emerged, I thought of the project, in part, as an academic, theoretical, and critical cover for what by then had become my daily practice of using "starling" form to try out in a different (poetic, lyrical, but also idiosyncratic and experimental) register the things I was engaged with critically, theoretically, and in academic contexts. In that sense, the seminar provided a kind of "fieldwork"—an investigation to test out hypotheses about the somewhat accidental, aleatory, and whimsical association of the starling with an experimental technology of writing, transcription, and the word. The results opened up a number of surprises, most of which stem from the thing that ought, perhaps, not to have come as a surprise—the fact that, although my starling form is indeed unique, there is nothing really original about it.[2] Most of the surprises, indeed, came from the discovery that a form of starling technology is deeply embedded, for theoretical, historical, and comparative cultural reasons, in various parallel formations: the formation of European lyric form, the formation of literary criticism and its dependence on quotation, the development of experimental technologies of avant-garde poetic and literary practice, and the formation of interdisciplinary study on the cutting edges of various theoretical trends (sound studies, interdisciplinary cognitive studies, animal studies).

Just about every poet seems compelled to write a poem about a starling.[3] And just about every critic seems to follow the starling's way of adapting the jargon of theory to his or her own idiom. One thing poet, critic, and starling all have in common is the habitual practice of quotation, citation, or allusion (mimicry, in a word). Whether conceived in terms of an alchemical process, a scientific method, or the molecular side effect of instinct, this habitual practice cannot be reduced, of course, to a single form. It deserves one, nonetheless, and this is what I have improvised: a technology for writing a short thought—a hypothesis, a sentence, what have you—in twenty-seven words composed in three columns of nine words (in groups of three) going from right to left.[4] For examples, see this Afterword's epigraph, the starling summaries that head each of the chapters in this volume, and the arrangement below of "Thirteen Starling Ways of Looking at a Blackbird" (which I turn to now to explicate starling form).

A word, to be sure, is an odd thing—too large and fuzzy to offer any kind of elemental linguistic particle, too small and ill-defined to provide the basis for the rhythm, the syntax, or meaning of something more. In a group of three words, the word may begin to make sense of its own arbitrary boundaries, whether as an auditory phenomenon (phoneme, syllable, rhythm, or rhyme) or visual cue (grapheme, lettering, spacing, or sign). Three times three words is magic enough to suggest a logical sequence and arbitrary enough to explore the limits of words. The break from one column of nine words to the next column of nine—like the break in the line of conventionally written lyric poetry—forces a turn in all senses. It is perhaps the most obvious reminder of the critical decision any reader always faces—the caesura already there in the poetic line (replicated every third word in starling form), the ambiguity of syntax that defines the prose sentence, the moment of cutoff when a reading encounters its own mimic-idiom (whether that of poet, critic, or starling). In nine words, a writer can probably quote enough of a poem to be recognizably quoting, citing, or mimicking. Consider the first line of the first of "Thirteen Starling Ways of Looking at a Blackbird": "among twenty some / snowy mountains the / only moving thing"—clearly an adaptation from the first two lines of Wallace Stevens's "Thirteen Ways of Looking at a Blackbird": "Among twenty snowy mountains / The only moving thing."

The way one reads is not always (maybe always not) the way one writes. Confronted with a square of twenty-some words organized three by three by three, from the start one has to decide how to read. Habit might dictate we read from left to right. So, to continue our blackbird example, we might initially read the first nine words as: "the only moving / was the eye / among twenty some"—or we might adjust to read in columns going down starting from the top left: "the only moving / thing this first / way of looking." The minute we recognize the quotations and allusions, we might stop reading altogether—smile, or frown, and turn away. But force of habit might also compel us to read in other ways, reread in older ways, follow other, or odder, or older habitual practices of reading writing—such, for example, as the way ancient Chinese or Japanese poetry has been written—going down in columns from right to left.

To follow the way Chinese or Japanese poetry was traditionally composed imposes a willful reversal of direction in the reading of an English word, line, and sentence. This might be considered either a characteristically "difficult" avant-garde maneuver, borrowing from the example of language poets or more recent experiments such as Susan Howe's "word squares," or it might be considered a throwback to older, Orientalized styles—or perhaps some combination of the two, like the mimicry of Stevens's own reduction of the first "blackbird" stanza to a haiku-like imagism, with the vivid movement of the written character of the Poundian Chinese word (the "only moving thing"). Here, the theory and the practice of writing—of prose and poetry—converge in the everyday material performance of critical argument and creative thought at the intersection of seemingly very distant and very different kinds of reading and writing systems. "Starling" form in this sense is a technique of transcription, translation (at an even more basic, molecular level, transliteration)—most evident in the adaptation of another creative or critical work (as with Stevens's "Thirteen Ways of Looking at a Blackbird" or my summaries of the articles in this volume).

Starling form is a technique of writing intended to submit the English word, line, and sentence to a daily exercise, a practicing, and a reflection on its limits. In this sense the reversal of direction imposed by following the traditional Chinese or Japanese way of writing in columns going down from right to left leads the writer to an experiment in transcription. Any ordinary sentence transcribed into the grid of three columns of twenty-seven words gets opened up to a reverse and perverse reading, revealing the kind of inner slips and accidents that poetry so often mines. What Giorgio Agamben claims about the effect of the poetic line—which forever suspends and defers the end of the poem—is something practically and theoretically discovered in the movement of writing and reading in starling form. Once one discovers that the poem can (or must) be read in at least the two directions set, on the one hand, by the conventional English line and, on the other hand, the traditional Chinese line, the reading and the writing submit themselves to this fuller reading. More often than one might expect, an English sentence conceals an inverted sense revealed in its transcription into starling form.

Wallace Stevens is in many ways the master of the art of this English-sentence inversion. The structure of "The Snow Man" might be the most obvious demonstration of this, all five stanzas following the movement of one sentence whose balance, syntactically, works as well backward and forward ("One must have a mind of winter" parallels "For the listener who listens in the snow"). What Stevens evokes with such pristine aesthetic economy is an ideal poem that, not unlike the ideal Chinese poem, might balance visually the yin and the yang of character, sound, and sense. This ideal balance is potentially there in any starling transcription (although it is the accidentals that are more revealing, creatively and critically, as we will discuss in a moment). So, for example, however difficult it may be to fit the fourth poem of Stevens's "Thirteen Ways of Looking at a Blackbird" into starling form, the exercise at least potentially reveals a balancing and reversal of contrasts in the poem's hierarchy of "a man and a woman" with the addition of the "blackbird." To be sure, the poem does not itself reverse the order to "a woman and a man." But starling transcription allows for the visual grasp of the pairing brought out by the third element (the "blackbird") that makes them "one": "a man and a woman are one . . ." "a woman and a man and a black bird."

Starling transcription, however, foregrounds accidental effects. Like translation, which cannot but change the semantic and semiotic effects of the words it translates, there are things that will need to be left out, or added, or changed. As with translation—indeed, following the same linguistic principles and on the assumption that even (and especially) English needs to be translated into English—starling transcription seeks to find and enhance the accidental effects and resources already there in the original. So, continuing with the example of the fourth poem of "Thirteen Ways," the first line of the original poem falls two words short of nine words, and one is faced with a decision—one might add the first two words of the next line ("a man") to produce this first (or last) starling column: "a man and a woman are one a man." This reveals the hierarchy of "a man and a woman" in the original poem. Alternatively, one might add something, as translators sometimes need to supplement a literal translation to make clearer sense of (or to draw out some ambiguity in) the original. My decision to add "thing unthinking" is a characteristic starling embellishment. It is the sort of thing that Mozart demonstrated in showing how his pet starling distorted the musical phrasing of his own work.[5] Only in the third column (or first column) of the transcription does it emerge that this added bit is itself an embellished mimicry of another, later Stevens poem, "Not Ideas about the Thing but the Thing Itself." It reads "That scrawny cry" as a reference not only to the bird call of that poem but as a reflection on the auditory effects of birds and things (and birds as things) throughout Stevens. It is the moment of starling transcription that tests the "starling" technology of Stevens's own "blackbird." As a poem itself, the starling transcription may be negligible. The act of transcription, nonetheless, makes it possible to read Stevens's "blackbird" as itself a starling—to hear the auditory effect of starling song in the visual effect of the "blackbird" that makes "a man and a woman" "one."

Each starling transcription of "Thirteen Ways of Looking at a Blackbird" might be read not only as a translation (from English to English). It might also be read as a kind of critical reading—either a citation from the stanza in question or a direct quotation accompanied by commentary (following either the parrots' way or the nightingales' way, according to Sarah Kay's argument), but more usually a combination between the two (the starling's way). The sequence of critical readings, citations, quotations, and commentary extends the argument developed in a different way in my critical essay (and its claims about the way Jorie Graham's poem prompts us to reread the work of Wallace Stevens). It is intended to demonstrate, in auditory and visual form, the way in which Stevens's "blackbird" might in fact already be a starling.

To call this "starling" form runs the risk of acknowledging, without attempting to correct, the Eurocentric bias of the very poetic and critical models it is designed to reimagine. Reading the vanishing trope of the starling in Jorie Graham—and the rereading of Wallace Stevens it prompts in turn—has led me to reassess the importance of that otherwise purely fortuitous and accidental naming. As with most such classificatory names, there is indeed a complex and comparative cross-cultural history attached to the trope of the starling's disappearance. As a matter of daily practice, the "starling" form is intended to provide a template for probing that sedimented history of translation. One of my earliest attempts to systematize "starling" form depended on imagining each column as a transcription into a different cultural framework—a different language, a different poetic tradition, even a different script. A working model of this is the starling heading for this Afterword ("cross-linguistic starling flight"), which meditates on the word for starling in English, Indonesian, and Chinese: starling, *jalak*, and 燕八哥 (yàn bā gē) (or simply 八哥 [bā gē]).

In light of my critical essay "Of Mimicry, Birds, and Words," I'm inclined to read this early effort as an instinctive, albeit not yet fully examined intuition about the three hypotheses I develop there concerning "starling" technology in a comparative cultural perspective. The Chinese [燕]八哥 ([yàn] bā gē), the Indonesian *jalak*, and the English "starling" all correspond to more or less the same bird, and each would likely be used to translate the other. Yet, as I discuss above in the context of the Indonesian term "jalak Bali," each term would likely conjure associations with a rather different kind of "starling" or "mynah": the Chinese crested mynah or starling, the Indonesian mynah (of various kinds), and the European starling.

Examples from Chinese, Indonesian, and English literature illustrate how each of these bird/word forms enacts a vanishing trope in a full literal and figurative sense. Consider three examples, each from very different literary-cultural contexts, but all of which might ultimately be interrelated: first, from *Hikayat Bayan Budiman* (Tale of the wise parrot), a fourteenth-century (1371) Malay version of an older Sanskrit version of "The Parrot's Tales"; second, from the celebrated eighteenth-century novel by Cao Xueqin, *The Story of the Stone*, also called *Dream of the Red Chamber* (c. 1760); and third, from a 1792 English translation of the *Tutinama* (a later version of the *Parrot's Tales* that is the basis for the Malay *Hikayat Bayan Budiman*). The Malay and the English texts both provide examples of what one might be tempted to call the primal scene of the starling's disappearance in world literature. Drawn from the ancient Sanskrit tradition of storytelling, both inaugurate the titular parrot's tales with the killing of the parrot's fellow bird—what else but a starling. The literal killing of the starling at the beginning of each tale is replicated by a resonant philological question about the bird's name. In the Malay version, the name is "Tiung" ("burung Tiung"), usually glossed in Indonesian with the generic term "jalak" (although it might be worth noting that the word "tiung" is also current in Indonesian as a term translated: "k.o. bird, mynah and other similar birds").[6] In the much later 1792 English translation of another version of these tales, translated as *Tales of a Parrot*, the text uses the word "sharyk" and glosses it accordingly: "a species of nightingale, which imitates the human voice in so surprising a manner, that if you do not see the bird you cannot help being deceived" (*Tales of a Parrot*, 5).

One reason to read these two versions as the primal scene of the starling's (dis)appearance in world literature is that there is a much more famous and more influential echo of both early on in the *Arabian Nights' Entertainments*, in "The Story of the Husband and the Parrot." A full reading of this episode would deserve at least an entire essay of its own in our volume (suffice it to say that the trick played on the parrot turns against the parrot its own mimetic powers and that here the single bird, standing for the paired birds that inaugurate the *Tūtī-nāma*, is killed within the space of the one episode). No trace of the starling remains in this version, though, enacting as a sort of trauma the appearance of the starling in the *Arabian Nights*, one of the most celebrated crossings over in world literature between different cultural traditions.[7]

The fleeting appearance of a starling in *The Story of the Stone* occurs in Chapter 41, a few chapters after the reader's discovery (in Chapter 35) that the main heroine, Dai-yu, possesses a rather gifted parrot able to mimic

Dai-yu's own interest in quoting and composing poetic lines. In the later chapter, the country-bumpkin grandmother, Grannie Liu, comments on the sophistication of everything she sees and hears in the Jia household, focusing her comments on a contrast between two kinds of pet birds, both of which impress her with their ability to talk:

> "Seems that in the city it isn't only the folks that are grander," she remarked. "The creatures too seem to be grander than what they are outside. Even the birds here are prettier, and they can talk."
>
> "What birds?" they asked her, curious.
>
> "I know the one on the golden perch on the verandah—him with the green feathers and red beak—is a polly parrot," she said defensively, "but that old black crow in the cage—he's grown a thingummy on his head and learned to talk, as well."
>
> The "crow" that she was referring to was a mynah. The others laughed at her mistake.
>
> (2:309)

The humor of the passage turns on a revealing misidentification of the bird Grannie Liu calls a crow. The translator, David Hawkes, helpfully explains that "the 'crow'" is really a "mynah"—and the reference to the "thingummy on his head" would seem to specify that this is a Chinese crested starling or mynah. Yet it is worth emphasizing that the original Chinese text does not give a name for the bird at all—between Grannie Liu's ignorance and the mocking knowledge of the Jia household the text leaves a silence where the starling makes its single appearance—all the more striking because it suggests that the starling has been there all along—an absent presence, even more enigmatic than the parrot, and indeed a kind of mute gloss on the parrot's own surprising appearance and disappearance as a figure able to mimic the characters' obsessive citation, quotation, and composition of lines of carefully composed classical Chinese verse.

Starling form is my deliberate, artificial, somewhat perverse, but nonetheless serious effort to make visible this disappearance of the starling within and across different literary formations—to theorize and to practice the technology of literary troping we see in these examples of the starling's disappearance in the translation across various different traditions of storytelling. I have repeatedly described this as a technology of the word. It might be equally as important to emphasize the crossing over of different media involved in the exercise of transliteration, transcription, or translation involved with starling form. The inherently ekphrastic features of the form, as I conceive and have attempted to practice it, might be illustrated with one final example.

Consider my attempt to transcribe in the most minimal way the painting that serves as a meditative center of attention for the volume *Mocking Bird Technologies* as a whole. The painting is entitled *Yogini with a Mynah Bird* (by the so-called Dublin Painter from Bijapur, early seventeenth century) and is discussed in Chapter 10 by Madeleine Brainered and Kaori Kitao. This painting is full of tropes of birds and flowers and designs, with numerous lyrical inscriptions (couplets on the outer right and left margins; on the outer top and bottom margins; and on the inner border on white ground, above and below). The figure of the woman is apparently described as "straight as an *alif*," suggesting the entire picture might be read as a meditation on its own calligraphic excess (there are also, "on reverse," verses by Katibi). Yogini and mynah bird form their own couplet, but the contextual relation of this pair to the literary traditions that surround it is not entirely clear. It seems to call for meditation on its own meditation. The mirroring of literary and cultural traditions concentrated in the mutual mirror gazing of yogini and mynah—and including the rich and complex cross-fertilization of different traditions projected onto the painting by the Metropolitan Museum of Art's catalogue (and by the other pieces chosen for the exhibit *Sultans of Deccan India*, including the images of parrots copied to and from European and Indian traditions of painting)—all this is concentrated, or stilled, by the singular look of the starling.

Strictly speaking, the "Dublin painter" has nothing to tell us about the starling form of "mocking bird technologies." In the enigmatic look and meditative silence of its mynah bird, this painting nonetheless exhibits a specimen that takes its shape and name, its visual and auditory tricks, from the bird and the word that is the vanishing point of cultural perspective across the world's mimetic traditions. Every starling appearance calls forth a starling manifesto.

Thirteen Starling Ways of Looking at a Blackbird (Transcriptions after Wallace Stevens)

I

the	was	among
only	the	twenty
moving	eye	some
thing	of	snowy
this	the	mountains
first	blackbird	the
way	a	only
of	starling	moving
looking	eye	thing

II

like	I	the
a	was	three
tree	looking	minds
on	at	of
which	all	the
starlings	three	tree
try	but	like
to	the	the
settle	mind	blackbirds

III

a	from	the
small	the	blackbird
part	world	whirled
a	of	in
starling	the	the
part	troupes	autumn
of	of	winds
the	the	a
pantomime	grackles	part

IV

the
thing
itself

a
scrawny
cry

thinking
the
starling

a
woman
and

a
man
and

a
black
bird

a
man
and

a
woman
are

one
thing
unthinking

V

when
the
starling

is
the
key

to
the
stillness

inflection
or
innuendo

the
blackbird
whistling

or
just
after

I
don't
know

which
to
prefer

the
beauty
of

VI

the
mood
traced

in
the
shadow

an
indecipherable
cause

the
shadow
of

the
starling
crossed

to
and
fro

icicles
once
filled

the
long
window

with
barbaric
glass

VII

walking
around
connecticut

at
the
feet

of
the
women

do
you
not

see
this
starling

how
the
blackbird

o
thin
men

why
do
you

imagine
golden
birds

VIII

in
knowing
what

starling
turns
it

takes
to
know

but
I
know

too
that
the

blackbird
is
involved

I
know
these

noble
accents
and

lucid
inescapable
rhythms

IX

of
one
starling

of
many
circles

of
one
blackbird

the
starling
marked

the
edge
it

marked
the
edge

when
the
blackbird

flew
the
starling

out
of
sight

X

even
the
birds

would
cry
out

sharply
the
sight

a
green
light

even
the
bawds

of
starling
euphony

at
the
sight

of
blackbirds
flying

in
a
green

XI

the
starling
shadow

of
his
equipage

for
black
birds

once
a
fear

pierced
him
in

that
he
mistook

he
rode
over

all
connecticut
in

a
glass
coach

XII

the
stern
particular

of
the
air

the
whitest
eye

the
black
bird

must
be
flying

somewhere
over
head

the
river
is

moving
over
stones

in
starling
stripes

XIII

<table>
<tr><td>the</td><td>in</td><td>it</td></tr>
<tr><td>blackbird</td><td>starling</td><td>was</td></tr>
<tr><td>sat</td><td>style</td><td>evening</td></tr>
<tr><td></td><td></td><td></td></tr>
<tr><td>in</td><td>and</td><td>all</td></tr>
<tr><td>the</td><td>it</td><td>after</td></tr>
<tr><td>cedar-limbs</td><td>was</td><td>noon</td></tr>
<tr><td></td><td></td><td></td></tr>
<tr><td>of</td><td>going</td><td>it</td></tr>
<tr><td>other</td><td>to</td><td>was</td></tr>
<tr><td>times</td><td>snow</td><td>snowing</td></tr>
</table>

NOTES

1. With the notable exception of my son, Keir GoGwilt, a violinist, who has adapted the form in his own practice.

2. Including the fact that the term “starling” has already been used to describe a poetic form—the Italian “stornello,” a folk verse form traced back to the fifteenth century (see *Princeton Encyclopedia of Poetry and Poetics*, 1360). “Starling” form might also be described as “unoriginal,” in the sense Marjorie Perloff develops in her account of the use of citation in twenty-first century avant-garde poetry in *Unoriginal Genius*.

3. A random sampling of great poems about starlings not discussed at all in the current volume: Paul Celan, “Ein Ring, zum Bogenspannen,” in *Breathturn into Timestead*; Edwin Morgan, “The Starlings of George Square”; Richard Wilbur, “An Event”; Robert Hayden, “A Plague of Starlings”; Sherman Alexie, “Avian Nights”; and Kathleen Jamie, “Wings over New York,” from *The Bonniest Companie*.

4. I have experimented using this form (1) as a form of lyric poetry (see my “In Defense of the French Horn,” 22); (2) as a way of citing passages in a more traditional form of literary criticism (my lecture on casement windows in *Jane Eyre* at Fordham, October 2012; and on the windows of *Heart of Darkness* at the New School, February 2016); (3) as a way of transliterating, transcribing, or translating words or things from one language, or medium, into another (e.g., Schubert songs, paintings, opera plots, poetry by Paul Celan); and (4) as accompaniment to musical performance (e.g., “Words and Music,” performance by Matt Aucoin and Keir GoGwilt, Peabody Essex Museum, March 7, 2014).

5. For a discussion of Mozart’s pet starling, see Meredith J. West and Andrew P. King, “Mozart’s Starling,” *American Scientist* 78 (March–April 1990): 106–114.

6. In Javanese, the word *tiyung* ("a certain bird") is considered a variant of *ciyung* ("a wild béyo [talking mynah bird]") and *siyung* (also "a wild béyo"), suggesting a repertoire of elemental questions of transcription in the writing of the name of a starling/mynah not yet caged. See Robson and Wibisono, *Javanese–English Dictionary*.

7. For a recent reassessment of the transcultural phenomenon of the *Arabian Nights*, see Srinivas Aravamudan, *Enlightenment Orientalism: Resisting the Rise of the Novel* (Chicago: University of Chicago Press, 2012).

WORKS CITED

Agamben, Giorgio. *The End of the Poem: Studies in Poetics*. Trans. Daniel Heller-Roazen. Stanford, Calif.: Stanford University Press, 1999.

Alexie, Sherman. *Face*. New York: Hanging Loose Press, 2009.

Aravamudan, Srinivas. *Enlightenment Orientalism: Resisting the Rise of the Novel*. Chicago: University of Chicago Press, 2012.

Bāna. *Princess Kādambarī*. The Clay Sanskrit Library. New York: NYU Press, 2009.

Cao, Xueqin. *The Story of the Stone*. Volume 2. Trans. David Hawkes. Harmondsworth: Penguin, 1977.

———. 紅樓夢: *A Dream of Red Mansions*. Volume 3. Trans. Yang Xianyi and Gladys Yang. Library of Chinese and English Classics. Beijing: Foreign Language Press, n.d.

Celan, Paul. *Breathturn into Timestead*. Trans. Pierre Joris. New York: FSG, 2014.

Cushman, Stephen, Clare Cavanagh, Jahan Ramazani, and Paul Rouzer, eds. *The Princeton Encyclopedia of Poetry and Poetics*. Princeton, N.J.: Princeton University Press, 2012.

GoGwilt, Christopher. "In Defense of the French Horn." *Beloit Poetry Journal* 60, no. 3 (2010).

Haidar, Navina Najat, and Marika Sardar. *Sultans of Deccan India, 1500–1700: Opulence and Fantasy*. Metropolitan Museum of Art. New Haven, Conn.: Yale University Press, 2015.

Hayden, Robert. *Collected Poems*. New York: Liveright, 1970.

Jamie, Kathleen. *The Bonniest Companie*. London: Picador, 2015.

Kay, Sarah. *Parrots and Nightingales: Troubadour Quotations and the Development of European Poetry*. University of Pennsylvania Press, 2013.

Morgan, Edwin. *Selected Poems*. Manchester: Carcanet, 2000.

Perloff, Marjorie. *Unorginal Genius: Poetry by Other Means in the New Century*. Chicago: University of Chicago Press, 2010.

Robson, Stuart, and Singgih Wibisono. *Javanese–English Dictionary*. Singapore: Periplus, 2002.

Tales of a Parrot; done into English from a Persian Manuscript, intitled Tooti Namêh, by a teacher of the Persic, Arabic, Hebrew, Syriac, Chaldaic, Greek, Latin, Italian, French and English Languages. [Trans. B. Gerrans.] London: Minerva, 1792.

Stevens, Wallace. *The Palm at the End of the Mind.* New York: Vintage, 1972.

West, Meredith, and Andrew P. King. "Mozart's Starling." *American Scientist* 78 (March/April 1990): 106–114.

Wilbur, Richard. *Collected Poems, 1943–2004.* New York: Harcourt, 2004.

Winstedt, R. O., ed. *Hikayat Bayan Budiman.* Kuala Lumpur: Oxford University Press, 1966.

CODA

Tornada, in Starling Form

Sarah Kay

stammering	clearing	the
and	the	troubadours
mute	lines	sing
scripted	for	a
by	whatever	bracing
starlings	sweet	air
linked	bird	through
and	sound	whistling
apart	leaves	branches

Mocking Bird Technologies is a fabulous volume. I love the innovative ways it engages with the repetitions integral to literary form and its constant "invitation for readers to develop their own critical and creative lines of inquiry across disparate fields of study," as GoGwilt and Holm write in the Introduction. Generously citing my earlier work on parroting, which looked at the functioning of quotations from troubadour songs, Christopher GoGwilt has asked me to contribute a *tornada* to the rich array of stanzas he and Melanie Holm have concatenated here. There are no exotic mynah or mocking birds in the troubadour corpus, but I was excited, reading *Mocking Bird Technologies*, to discover that "starling form," initially seen by GoGwilt as an altogether idiosyncratic invention of his own, was already appearing in twelfth-century Occitania and that it was, as he puts it in his Afterword, "deeply embedded . . . in . . . the formation of European lyric form, the formation of literary criticism and its dependence on quotation." In reprising here the concept of starling form I follow the conventions of the medieval *tornada*, which echoes, starling-like, the song it concludes, and in retrospectively uncovering the starlings' formal insinuation within medieval poetics I hope to contribute a further strand to the many "critical and creative lines of inquiry" that traverse this volume.

A number of troubadour songs survive whose versification may be set out either in compact stanzas with relatively long lines punctuated by internal rhymes or in long straggling stanzas made up of shorter and very short lines. Appel prints Arnaut Daniel's famous "L'aur' amara" in stanzas of seven eight- and ten-syllable lines, each of which is divided by at least one internal rhyme, marked typographically by an extended space. None of the end-rhymes nor the internal rhymes have an exact partner within the stanza, only from one stanza to another, as quotation of the first two stanzas of his edition shows:

L'aur' amara   fa·ls bruels brancutz
clarzir,   que·l dous' espeys' ab fuelhs,
e·ls letz   becx   dels auzels ramencx
4 te balbs e mutz   pars   e non-pars.
Per qu'ieu m'esfortz   de far e dir   plazers
a manhs? Per ley   qui m'a virat bas d'aut,
7 don tem morir,   si·ls afans no·m asoma.
[The bitter breeze   makes the branchy thickets
grow clear   that the sweet [breeze] makes thick with leaves
and the merry   beaks   of the birds on the bough
it holds stammering and mute,   partnered   or unpartnered.
Why do I strive   to compose and say   pleasing things
to many? Because of her   who has cast me low from high,
for whom I fear I will die   if she doesn't end my torments.

'Tan fo clara   ma prima lutz
d'eslir   lieys, don cre·l cors los huelhs,
non pretz   necx   mans dos angovencs
11 d'autra. S'eslutz   rars   mos preyars
pero deportz   m'es d'auzir   volers
bos motz ses grey   de lieys, don tan m'azaut
14 qu'al sieu servir   sui de pe tro qu'al coma.
[So clear   was the first bright light
by which I chose   her for whom my heart believes my eyes
that I don't value   secret   summons at two pence
from another lady. Gleams forth   but rarely   my entreaty
but it is a pleasure   to me to hear   the wishes,
good words without rancor,   from her who pleases me so much
that at her service   I am from head to foot.]

(Appel, *Provenzalische Chrestomathie*, 66; translation mine)

Other editors, however, beginning with Ugo Angelo Canello in 1885, have opted to arrange the text in seventeen-line stanzas whose lines vary in length from one to six syllables and are loosely linked by a scattering of end-rhymes (here line 2 rhymes with line 9, and 11 with 12; later *auzir* rhymes with *servir*):

> L'aura amara
> fa.ls bruoills brancutz
> clarzir
> que.l doutz espeissa ab fuelhs,
> e.ls letz
> becx
> dels auzels ramencx
> te balps e mutz
> pars
> e non-pars . . . (*etc.*)

(Canello, *La vita e le oper del trovatore Arnaldo Daniello*, Song IX, 105)

This is the format now universally adopted for this song.[1] But *Mocking Bird Technologies* suggests that Canello, even as he innovated by launching this astonishingly etiolated strophe, also linearized Arnaut's complexly sprung versification and thereby limited its creative potential. Under its inspiration I wish to explore instead the multiple alternative readings enabled by the seven-line format, set out on the model of those lines that are divided by internal rhyme into three parts, so that the whole text is arranged in three columns, starling style:

L'aur' am**ara**	fa·ls bruels	branc**utz**
clarz**ir**,	que·l dous' espeys'	ab f**uelhs**,
e·ls l**etz**	b**ecx**	dels auzels ram**enc**x
te balbs e m**utz**	p**ars**	e non-p**ars** . . . [etc.][2]

The resulting texture is lumpier than GoGwilt's elegant minuet of single words. But medieval scribes, with their scant regard for word boundaries, leave readers of the troubadours free to follow different norms. The medieval text, too, rejoices in the mocking bird property of rhyme. In the passage above, syllables that rhyme within or between stanzas are marked in bold; note how very few word clusters are *not* bounded by rhyme. The song's music has not survived, but typically melodic cadences would follow the contours of the longer lines, with performers having the freedom to create internal rhythms of their own that could further enhance the unity of these clusters. In any event, even if this instance of starling form departs from the examples in *Mocking Bird Technologies*, some of the same effects—what GoGwilt in the Afterword to this volume calls "technologies of the word" "between poetics and theory"—result from its deployment in Arnaut's song.

The bitter breeze, reading vertically down the first column, both brightens and silences, stripping out sound and color; the thickets thicken with conjoined beaks as we read down the second; isolated birds are scattered, in the third, among reiterations of branches and greenery. Reading diagonally, *amara* assonates with *pars* just as *mutz* rhymes with *brancutz*. The alphabetical underlay of the opening phrases leaps into prominence (**a**ura **a**mara—**b**ruels **b**rancutz—**cl**arzir—**d**ous'—**e**speys—**f**uelhs). The cipher-like technology of Arnaut Daniel's poetry, fully exposed, both invites and obstructs interpretation. His rhythm of short, seemingly transposable units within larger, interactive groupings that come together to form a mobile "stanza" presents a recognizably starling form whose "stammering" or even "mute" style is captured reflexively by its diegetic birds. Mimicked by his form, the stammering of the birds will also be imitated by the poet who refers in stanza II to his rarely voiced entreaties.

There are doubtless other troubadour songs that likewise attest this form. I shall consider just one, "Cars, douz" by the troubadour Raimbaut d'Aurenga, because Arnaut's birds that stammer in the thickets may well have been prompted by Raimbaut's wrens calling back and forth among the cork oaks and the crickets. Arnaut's poetics were clearly much influenced by the troubadour from Orange, and there are considerable formal similarities between the two songs that chime with the modeling, in both, of an avian poetic that at once defies and demands imitation by the human voice.

Notoriously difficult, "Cars, douz" celebrates its own obscurity through repetition of the internal refrain-word *car*, which occurs in the same position at the beginning of every stanza. The word appears to evoke what was known by the troubadours as *trobar car*, understood as a form of poetic composition (*trobar*) whose charm, value, and preciosity are reflected in the shimmering equivocations of *car* ("dear," "rare," "precious"). Although the song contains several further internal rhymes, and therefore could be set out in the long, thin, straggling strophes that Canello accords Arnaut, the song is usually printed in compact stanzas of nine eight- or seven-syllable lines. As with some lines of "L'aur' amara," an effect of the internal rhymes is to divide the first line of "Cars, douz" into three; in arranging the song's first stanza below, I have extended this pattern to all the other lines and also grouped the lines three by three, starling style. Again as with Arnaut, the resulting text typically clumps groups of words together, rather than ringing the changes on isolated words like GoGwilt's examples, but many of these clusters are bounded by some form or other of rhyme.[3] Alas, once again we do not have any musical notation and can only speculate how this human imitation of the wren's song might sound.

Cars,	douz e f**einz**	de beder**esc**
m'es	sos bas ch**anz**,	vas cui m'az**erc**;
c'ap joi	s'esp**an**,	viu e n**oire**,
el tems	qe gr**ill**	prob del **siure**
chanton	el m**ur**	jos lo **caire**,
qe·s	compassa	e s'esq**aira**,
sa vos	cha plus leu	qe **siura**,
e ja	nuls non	s'i as**erga**
mas grils	et la	beder**esca**.
[Precious,	sweet and inventive	is the wren's
low song	to me	towards which I am exalted,
for with joy	it pours forth,	lives and teaches [*or* nourishes],
at the time	when crickets	beside the cork oak
are singing	in the wall	under the dressed stone,
for it [the wren's song]	is measured	and squared [i.e., dressed, like stone],
its voice	falls more lightly	than cork,
and never	may any	rise up with it
except a cricket	and the	female wren.]

Unlike most troubadour songs, though similar to "L'aur' amara," "Cars, douz" contains no rhymes within the stanza, only between one stanza and another; in the text above, these rhymes are picked out in bold. "Cars, douz" is yet more frugal in in its use of rhyme than "L'aur' amara," but it does deploy a variant technology, so-called derived rhyme, in which different morphological forms of the same word are used at the rhyme: In this stanza, the words concerned are *bederesc–bederesca*, *azerc–aserga*, *siure–siura*, and *caire–esqaira*. The lines thus linked form an extended chiasmus with only the third line, ending in *-oire*, lacking a corresponding derived form *-oira*. Thus line 1 is paired with line 9, line 2 with line 8, line 4 with line 7, and line 5 with line 6. Analogously to how the stanzas are conjoined by the same initial word *car*, these line-final words "rhyme" on their opening syllables, not their closing ones, so that the song is bound together by differently rhythmed repetitions of both initial and final sounds. In a tour de force of virtuosity, the composition maintains the same rhyme sounds and rhyme scheme through seven full stanzas and two *tornadas*.

A consequence of the challenging rhyme scheme is that the semantically heaviest words, whether they evoke wrens, cork oaks, or masonry, are all clustered in the third of the three columns. Ideas of singing, voice, and song are reiterated through the first as desirable if not always attainable, while the middle column contains the largest proportion of adjectives and verbs, presenting the singing voice as pleasing, creative, low-key, and precise. But the starling form also potentiates significant movements along diagonal axes, as when *cars* (line 1) encounters *leu* (line 7), since *leu*, "light, easy," is often associated with an accessible style that is poetically quite the contrary of *car*. Another interesting diagonal is the coincidence between the crickets (*grill*) in line 4 and the cricket (*grils*) in line 9, which causes the latter to masquerade briefly as a rhyme word. Starling form makes it possible to see that whereas a *car* poetics of pleasure is at stake in the first column, and the possibility of a *leu* poetics is raised in the second, the full challenge of incorporating rare (*car*) words into the text is faced in column three.

Over the course of the stanza, then, the connotations of *car* veer off in competing directions, as do the values associated with the wren's song. Its voice combines the lightness of cork with the precise measurement and craftsmanship of stones dressed by a stonemason, stones whose implied size and weight contrast with both cork's soft, airy texture and the bird's soft, tiny body. The wren's song is low—or even quiet (*bas*)—yet it is widely audible (*s'espan*), alert or alive (*viu*), and formative (*noire*). Occupying the place of the rhyme with no derived partner, the word *noire* affirms the lone rhyme's productive role, since *norir* means "feed," "bring up, rear," "educate," and even "act as patron to."[4] The poet has become the nursling or protégé of the wren's song; aspiring to raise himself up to it,[5] even if the singer has to concede in the end that only the female wren, and perhaps the odd chirruping cricket, can rise to its challenge. The wren's song is *car*: endearing, precious, refined, rare to the point of being barely reproducible, yet the object of emulation.

Marc M. Vuijlsteke has cited this stanza as exemplifying a widespread troubadour topos: that of positioning the song that is sung as aspiring, but failing, to reproduce a postulated ideal song.[6] Raimbaut's wren, trilling in a low voice as it hops around the stone wall, inaugurates a comic admission of the impossibility for the instance to equate to the exemplar. Raimbaut's inability to measure up to a wren, even though admittedly a wren that measures up to monumental masonry, is as inglorious as Arnaut's stammering like the birds in the wintery branches. Both songs can be read more productively in the framework that GoGwilt terms the "starling effect," which repositions human language, education, poetry, and technology within a broader problematization of nature-culture.[7] The troubadour songs considered in this *tornada*—and many others—blend aspiration with derision and virtuosity with ruefulness, as humans concede their performance may do no more than inadequately imitate avian technologies. Masterworks of poetic resource and essays in poetics, these songs are also critical engagements with the aesthetic as such, as founded in interspecies mimicry.

NOTES

1. In quoting from Canello's edition I have followed his spelling and punctuation, which differ from Appel's. The text continues to figure as Song IX in subsequent editions (see, for example, the standard editions by Toja, Perugi, Eusebi, and Wilhelm).

2. I have retained Appel's spellings in order to facilitate reference to my translation of these lines.

3. Aside from the emboldening of rhyme syllables and the starling form layout, both introduced by myself, this is the text published by Paterson in her *Troubadours and Eloquence*, 147. Paterson bases herself with few changes on the edition of Marshall, "On the Text and Translation of a Poem by Raimbaut d'Orange." The translation is my own, with thanks to Marshall and Paterson, and also to Vuijlsteke, "Vers une lecture de *Cars, douz e fenhz* de Raimbaut d'Orange."

4. Hackney has exploited the generative power of this particular rhyme to a different end in "Privy to the Game." In this and two other related troubadour songs, Hackney discerns a micropoem built from the accumulated instances of the *rim estramp* in each parent song.

5. Vuijlsteke, "Vers une lecture," 514, translates this as "le bas chant du roitelet vers lequel je m'élève."

6. Vuijlsteke, "Le critique moderne et la lyrique médiévale des troubadours provençaux," 111: "Le chant du *bederesc* est le modèle idéal auquel tente de se référer le poète parce que, d'une part, c'est le *Joy* (résultant de la stricte observance de la *fin'Amor*, le code de l'éthique courtoise) qui en est le principe vital (*c'ap Joi s'espan, viu e noire*) et que, d'autre part, il est constitué selon toutes les règles de l'art (*qu' s compassa e s'esqaira*): Réunissant en lui les deux composantes nécessaires de la *canso* idéale—conformité au code éthique et au code poétique—il peut en effet servir de modèle au poète et devenir ainsi le chant *vas cui met que, d'aazerc*" (111).

7. GoGwilt, in his contribution to this volume "Of Mimicry, Birds, and Words," writes: "This starling effect repositions the comparative study of

language, the aesthetic education of humankind, and the biopolitical apparatus of modern technology in the most general of senses within a global constellation of different and overlapping systems of both nature and culture."

WORKS CITED

Appel, Karl. *Provenzalische Chrestomathie.* Leipzig: O. R. Reisland, 1895.

Canello, Ugo Angelo. *La vita e le oper del trovatore Arnaldo Daniello.* Halle: Max Niemeyer, 1883.

Hackney, Melanie A. "Privy to the Game: Three Generations of Troubadour Micropoetry." *Tenso* 31, nos. 1/2 (2016): 1–42.

Marshall, J. H. "On the Text and Translation of a Poem by Raimbaut d'Orange (*Cars douz*; ed. Pattison, I)." *Medium Aevum* 37, no. 1 (1968): 12–36.

Paterson, Linda M. *Troubadours and Eloquence.* Oxford: Clarendon, 1975.

Vuijlsteke, Marc M. "Le critique moderne et la lyrique médiévale des troubadours provençaux." *Philosophie* 38, no. 2 (1986): 107–116.

———. "Vers une lecture de *Cars, douz e fenhz* de Raimbaut d'Orange (P-C, 389, 22). Notes et commentaires sur l'édition de J. H. Marshall." In *Études de philologie romane et d'histoire littéraire offertes à Jules Horrent à l'occasion de son soixantieème anniversaire*, ed. Jean Marie d'Heur and Nicoletta Cherubini, 509–516. Liège: n.p., 1980.

ACKNOWLEDGMENTS

We would like to thank Tom Lay, Richard Morrison, and Eric Newman, along with the whole team at Fordham University Press, for their dedication, support, and vision in seeing this project through from beginning to end. We are also very grateful for the criticism and insight of the two anonymous reviewers. Chris thanks all his family and friends for their encouragement and support in helping stage starling forms of various kinds. Melanie would like to thank her little starlings, Willow and Heath, for filling her world with surprise and delight.

CONTRIBUTORS

MADELEINE BRAINERD taught at Washington University in St. Louis and at Excelsior College. Since 2004 she has taught therapeutic yoga and medical qi gong in New York City, at the Integral Yoga Institute, Kenshikai Dojo, Gouverneur Hospital, and other venues. She studies histories of yoga's intersections with ecological in/justice, animality, and affect theory.

JOE CONWAY is an Assistant Professor of American Literature at the University of Alabama in Huntsville. His articles have appeared or are scheduled to appear in the journals *Women's Studies*, *Early American Literature*, and *Nineteenth-Century Contexts*. He is currently at work on a monograph about the social life of antebellum money that charts how discourses of noneconomic phenomena such as medicine, race, nationalism, and aesthetics informed nineteenth-century debates about what constitutes good money.

FRASER EASTON is Associate Professor of English, University of Waterloo, Canada. A specialist in eighteenth-century literature, he has published on Jane Austen, Daniel Defoe, Maria Edgeworth, and Christopher Smart, as well as on newspaper records and historical accounts of passing women in the eighteenth century.

CHRISTOPHER GOGWILT is Professor of English and Comparative Literature at Fordham University. He is the author of *The Passage of Literature: Genealogies of Modernism in Conrad, Rhys, and Pramoedya* (Oxford, 2011), *The Fiction of Geopolitics: Afterimages of Culture from Wilkie Collins to Alfred Hitchcock* (Stanford, 2000), and *The Invention of the West: Joseph Conrad and the Double-Mapping of Europe and Empire* (Stanford, 1995).

SHARI GOLDBERG is Assistant Professor of English at Franklin & Marshall College in Lancaster, Pennsylvania. She is the author of *Quiet Testimony: A Theory of Witnessing from Nineteenth-Century American Literature* (Fordham, 2013). She has also published essays on silence, politics, and personhood in American literature. Her current research focuses on late-nineteenth-century models of mind and person in narrative and psychological writing.

MELANIE D. HOLM is Assistant Professor of the English Department and Graduate Program of Literature and Criticism at Indiana University of Pennsylvania. She also teaches in the university's Women's and Gender Studies program. Her scholarly focus is on eighteenth-century literature and skepticism.

SARAH KAY teaches French and Medieval Studies at New York University. She has written widely on medieval literature across languages, genres, and periods; her work combines the study of medieval texts, especially troubadour songs, with philosophical and theoretical inquiry. Her two most recent books are *Parrots and Nightingales: Troubadour Quotations and the Development of European Poetry* (2013) and *Animal Skins and the Reading Self in Medieval Latin and French Bestiaries* (2017).

KAORI KITAO (William R. Kenan Jr. Professor of Art History, Emerita, Swarthmore College) taught art history at Swarthmore College from 1966 to 2001. She was born in Tokyo and studied architecture at UC Berkeley and art history at Harvard. Her main specialization is Italian renaissance and baroque art; she has also taught courses in cinema history, material culture, urban studies, and Japanese architecture.

HOLT V. MEYER is Professor of Slavic Studies at Erfurt University. He is the author of *Romantische Orientierung* (1995) and numerous articles and has co-edited the collections *Juden und Judentum in Literatur und Film des slavischen Sprachraumes. Die geniale Epoche* (1999), *Inventing Slavia* (2005), *Schiller: Gedenken—Vergessen—Lesen* (2010), and *Gagarin als Archivkörper und Erinnerungsfigur* (2014). He is co-editor of the new book series Spatio-Temporality. Practices—Concepts—Media (De Gruyter). He is currently working on a book about the official Stalinist Pushkin celebrations of 1949.

ISABEL (ANNIE) MOORE completed her Ph.D. in comparative literature at the University of California–Irvine. From 2011 to 2013, she held a postdoctoral fellowship in English at the University of Victoria. She has published on contemporary Irish and Canadian poetry, and her book project is titled *The Ends of Lyric Life: A Theory of Biopoetics.*

FAWZIA MUSTAFA is Professor of English and African and African American Studies at Fordham University. She also teaches in the university's Comparative Literature and Women's, Gender and Sexuality Studies Programs. The author of *V. S. Naipaul* (1995), she has published numerous articles on postcolonial literature and development.

GAVIN SOURGEN is Visiting Assistant Professor of English at the Harriet L. Wilkes Honors College of Florida Atlantic University. He completed his D.Phil. at Balliol College (Oxford) in 2013, concentrating on the transitional poetics of Lord Byron's verse, and has published on Byron, Coleridge, and romantic aesthetics in general.

INDEX